STUDENT'S SOLUTIONS MANUAL

DISCRETE MATHEMATICS

FIFTH EDITION

John A. Dossey

Illinois State University

Albert D. Otto

Illinois State University

Lawrence E. Spence

Illinois State University

Charles Vanden Eynden

Illinois State University

PEARSON

Addison
Wesley

Boston   San Francisco   New York
London   Toronto   Sydney   Tokyo   Singapore   Madrid
Mexico City   Munich   Paris   Cape Town   Hong Kong   Montreal

Reproduced by Pearson Addison-Wesley from electronic files supplied by the author.

Copyright © 2006 Pearson Education, Inc.
Publishing as Pearson Addison-Wesley, 75 Arlington Street, Boston, MA 02116.

ISBN   0-321-30517-5

Contents

Chapter 1

An Introduction to Combinatorial Problems and Techniques

1.1 THE TIME TO COMPLETE A PROJECT

1. The total project time is 33, and the two critical paths are A-B-D-F-G and A-C-E-F-G, as shown in the figure below.

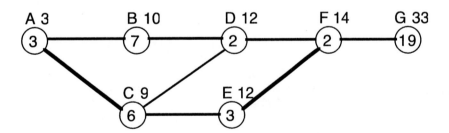

1

3. The total project time is 43, and the critical path is B-D-E-G, as shown in the figure below.

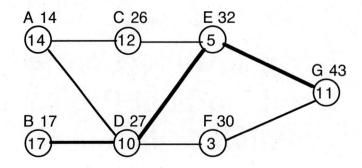

5. The total project time is 20.7, and the critical path is A-D-H-K, as shown in the figure below.

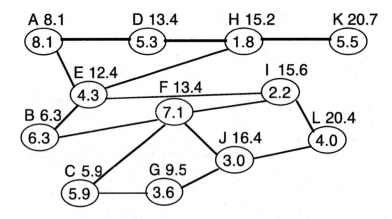

7. The total project time is 2.1, and the critical path is A-C-E-H-J, as shown in the figure below.

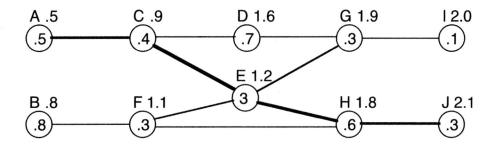

9. The total project time is 23, and the critical path is A-D-F-G, as shown in the figure below.

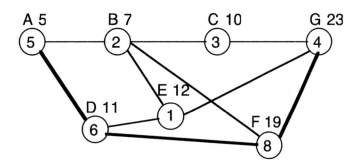

11. The total project time is 24, and the critical path is B-C-F-G, as shown in the figure below.

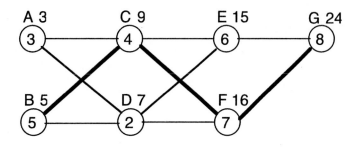

13. The total project time is 15.7, and the critical path is D-I, as shown in the figure below.

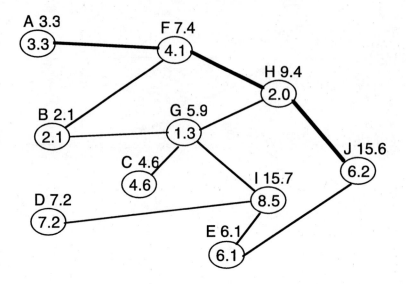

15. The total project time is 0.29, and the critical path is E-F-C-D-I, as shown in the figure below.

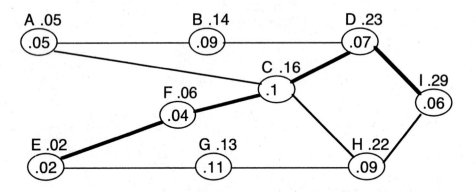

17. The total project time is 27 minutes, as shown in the figure below.

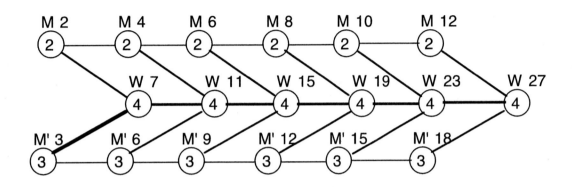

19. The total project time is 15 days, as shown in the figure below.

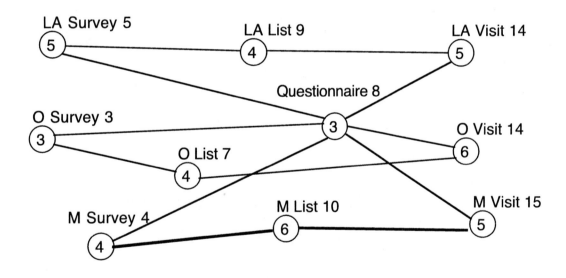

1.2 A MATCHING PROBLEM

1. $5! = 5 \cdot 4 \cdot 3 \cdot 2 \cdot 1 = 120$

3. $8!/3! = 8 \cdot 7 \cdot 6 \cdot 5 \cdot 4 = 6720$

5. $8!/2!6! = 8 \cdot 7/2 = 28$

7. $P(7,4) = 7 \cdot 6 \cdot 5 \cdot 4 = 840$

9. $P(10,7) = 10 \cdot 9 \cdot 8 \cdot 7 \cdot 6 \cdot 5 \cdot 4 = 604{,}800$

11. $P(9,4)/5! = 9 \cdot 8 \cdot 7 \cdot 6/120 = 25.2$

13. $P(6,6) = 6! = 720$

15. $P(8,3)/P(3,3) = 8 \cdot 7 \cdot 6/3! = 56$

17. $P(9,9) = 9! = 362{,}880$

19. $P(6,6) = 720$

21. Since he can choose one of 5 coats or no coat, one of 4 pairs of slacks, one of 6 shirts, and a tie or no tie, the answer is $6 \cdot 4 \cdot 6 \cdot 2 = 288$.

23. $P(7,3) = 7 \cdot 6 \cdot 5 = 210$

25. $P(5,3) = 5 \cdot 4 \cdot 3 = 60$

27. $P(9,6)P(11,6) = 60480 \cdot 332640 = 20{,}118{,}067{,}200$

29. Since we can turn the ring over we divide by 2, and since any key can be the first key we divide by 6. The answer is $6!/(2 \cdot 6) = 60$.

31. We have $P(n, n-1) = n!/(n - (n-1))! = n!/1! = n!$.

1.3   A KNAPSACK PROBLEM

1. False, since $1 \in A$, but $1 \notin B$.

3. False; 2 is not a set.

5. False; 10^6 is not odd.

7. True.

9. False; $2 \neq C$ and $2 \neq E$.

11. False; the set has only 3 elements.

13. False; $\varnothing \neq 1$ and $\varnothing \neq 2$.

15. $264 + 188 + 65 + 25 + 170 + 22 = 734 > 700$, so the selection is not acceptable.

17. $264 + 203 + 7 + 92 + 25 + 80 = 671 \leq 700$, so the selection is acceptable. Its total rating is $9 + 8 + 6 + 2 + 3 + 7 = 35$.

19. The ratios are approximately .139, .034, .032, .039, .077, .857, .022, .123, .120, .035, .088, and .182. The experiments chosen (in order) are 6, 12, 1, 8, 9, 11, 5, 4, and 7, for a total weight of $7 + 22 + 36 + 65 + 25 + 80 + 104 + 203 + 92 = 634$ kilograms. The total rating is $6 + 4 + 5 + 8 + 3 + 7 + 8 + 8 + 2 = 51$.

21. $2^7 = 128$

23. $2^5 = 32$

25. If $n = m + 1$ there are 2 (m and $m + 1$), if $n = m + 2$ there are 3 (m, $m + 1$, and $m + 2$), and in general if $n = m + k$ there are $k + 1 = (n - m) + 1$.

27. $2^5 - 1 = 31$

29. $(2^{40}$ sets$)(1$ sec$/106$ sets$)(1$ hr$/3600$ sec$)(1$ day$/24$ hr$) = 12.7$ days

1.4 ALGORITHMS AND THEIR EFFICIENCY

1. Yes; the degree is 2 because the highest power of x is 2.

3. No, because of the $1/x^2$ term.

5. No, because of the division.

7. Polynomial

S
3
$3 + 5(2) = 13$

Horner's

S
5
$3 + 2(5) + 3 = 13$

9. Polynomial

S
-7
$-7 + 5(2) = 3$
$3 + 2(2^2) = 11$
$11 - 2^3 = 3$

Horner's

S
-1
$2(-1) + 2 = 0$
$2(0) + 5 = 5$
$2(5) - 7 = 3$

11. $110101 \rightarrow 110111 \rightarrow 110110$

13. $001101 \rightarrow 001111 \rightarrow 001110$

15.

k	j	a_1	a_2	a_3
3		1	0	1
2		1	1	1
	3	1	1	0

17.

k	j	a_1	a_2	a_3	a_4
4		1	1	0	1
3		1	1	1	1
	4	1	1	1	0

19. See answer in back of book.

21. See answer in back of book.

23. (3^{20} operations)(1 sec/10^6 operations)(1 min/60 sec) = 58 minutes

($100 \cdot 20^3$ operations)(1 sec/10^6 operations) = .8 seconds

25. (3^{40})(1/10^6)(1/3600)(1/365) = 385,517 years

($100 \cdot 403$)(1/106) = 6.4 seconds

27. There are $n + 1$ comparisons, n additions of 1, and n multiplications by k, for a total of $3n + 1$ operations.

29. There are 1 multiplication (in step 1), n comparisons (beginning step 2), $n-1$ additions (step 2(a)), $n - 1$ multiplications (step 2(b)), and $n - 1$ additions (step 2(c)), for a total of $4n - 2$ operations.

31.

1	S
1	-7
2	$-7 + 2 \cdot 5 = 3$
4	$3 + 4 \cdot 2 = 11$
8	$11 - 8 = 3$

33. There are $n + 1$ comparisons (beginning step 2), n multiplications (step 2(a)), n additions and n multiplications (step 2(b)), and n additions (step 2(c)), for a total of $5n + 1$ operations.

SUPPLEMENTARY EXERCISES

1. The total project time is 18, and the critical path is B-D-G-I, as shown in the figure below.

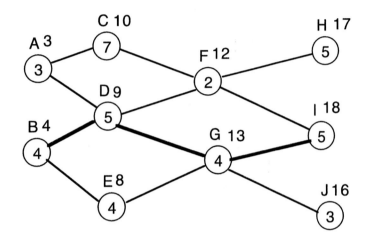

3. The total time to make and box 5 toys is 28 minutes, as shown in the figure below.

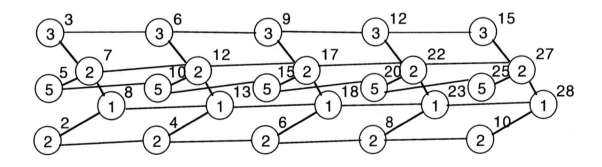

5. $11 \cdot 10 \cdot 9 \cdot 8 \cdot 7 \cdot 6 = 332{,}640$

7. $11 \cdot 10 \cdot 9 = 990$

9. False, since $10 \in B$ but $10 \notin C$.

11. False, since 6 is not a set.

13. True. Every element of $\{1\}$ (namely 1) is an element of A.

15. True. Every element of $\varnothing$ (there are none) is an element of B.

17. $2^5 = 32$

19. The spokesperson can be chosen in 5 ways, and the remaining students, if any, in $2^4 = 16$ ways. Thus there are $5 \cdot 16 = 80$ ways.

21. Yes, and its degree is 100, the highest power of x.

23. No, because the exponent 1.5 is not a nonnegative integer.

25.

$$\frac{S}{3}$$

$$3 \cdot 3 + 0 = 9$$
$$3 \cdot 9 + 4 = 31$$
$$3 \cdot 31 - 5 = 88$$

27. See the answer in the back of the book.

29. The values of s are 0, $0 + 1 = 1$, $1 + 2 = 3$, $3 + 3 = 6$, $6 + 6 = 12$, $12 + 9 = 21$, $21 + 18 = 39$.

31. There will be n comparisons at the beginning of step 2, $n - 1$ additions and $n - 1$ multiplications in step 2(a), and $n - 1$ additions in step 2(b), for a total of $4n - 3$ operations.

Chapter 2

Sets, Relations, and Functions

2.1 SET OPERATIONS

1. The union of A and B contains all of the elements in A or B or both; so

$$A \cup B = \{1, 2, 3, 4, 5, 6, 7, 8, 9\}.$$

The intersection of A and B contains all of the elements common to both A and B; so

$$A \cap B = \{3, 5\}.$$

The difference of A and B consists of the elements in A that are not in B; so

$$A - B = \{2, 7, 8\}.$$

The complement of a set X contains all of the elements in the universal set U that are not in X; so

$$\overline{A} = \{1, 4, 6, 9\} \qquad \text{and} \qquad \overline{B} = \{2, 7, 8\}.$$

3. As in Exercise 1, we have $A \cup B = \{1, 2, 3, 4, 7, 8, 9\}$, $A \cap B = \varnothing$, $A - B = \{1, 2, 4, 8, 9\}$, $\overline{A} = \{3, 5, 6, 7\}$, and $\overline{B} = \{1, 2, 4, 5, 6, 8, 9\}$.

5. The set $A \times B$ contains all the ordered pairs (a, b), where $a \in A$ and $b \in B$. Thus

$$A \times B = \{(1,7), (1,8), (2,7), (2,8), (3,7), (3,8), (4,7), (4,8)\}.$$

7. As in Exercise 7, $A \times B = \{(a,x), (a,y), (a,z), (e,x), (e,y), (e,z)\}.$

9. Draw a Venn diagram for $\overline{B}$ and then one for $A \cap \overline{B}$. From the latter, we obtain the following diagram for $\overline{(A \cap \overline{B})}$.

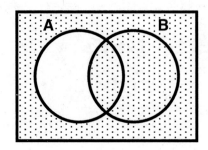

11. First draw a Venn diagram for $B \cup C$. From this we obtain the following diagram for $\overline{A} \cap (B \cup C)$.

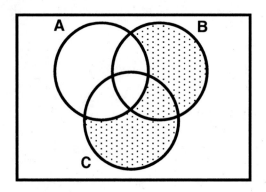

13. One possible example is: $A = \{1\}$, $B = \{2\}$, and $C = \{1, 2\}$. Then $A \cup C$ and $B \cup C$ both equal C, but $A \neq B$.

15. One possible example is: $A = \{1, 2\}$, $B = \{1, 3\}$, and $C = \{2, 3\}$. Then $A - C$ and $B - C$ both equal $\{1\}$, but $A \neq B$.

17. By Theorem 2.1, $A \cap (B - A) = A \cap (B \cap \overline{A}) \subseteq A \cap \overline{A} = \varnothing$. So $A \cap (B - A) = \varnothing$.

19. We have $(A - B) \cap (A \cup B) = A - B$ because $A - B \subseteq A \subseteq A \cup B$.

21. We have $\overline{A} \cap (A \cup B) = (\overline{A} \cap A) \cup (\overline{A} \cap B) = \varnothing \cup (B \cap \overline{A}) = B \cap \overline{A} = B - A$.

23. We have $A \cap \overline{(A \cap B)} = A \cap (\overline{A} \cup \overline{B}) = (A \cap \overline{A}) \cup (A \cap \overline{B}) = \varnothing \cup (A \cap \overline{B}) = A \cap \overline{B}$.

12

25. The set $A \times B$ contains all the ordered pairs (a, b), where $a \in A$ and $b \in B$. Since there are m possible choices of a and n possible choices of b, the multiplication principle shows that there are mn possible ordered pairs (a, b).

27. The equality $A \cup B = A$ holds if and only if $B \subseteq A$.

29. Let $x \in A \cup (B \cap C)$. Then $x \in A$ or $x \in B \cap C$. So $x \in A$, or $x \in B$ and $x \in C$. Thus $x \in A$ or $x \in B$, and $x \in A$ or $x \in C$. Hence $x \in A \cup B$ and $x \in A \cup C$. Therefore $x \in (A \cup B) \cap (A \cup C)$. This argument shows that

$$A \cup (B \cap C) \subseteq (A \cup B) \cap (A \cup C).$$

Reversing the steps above proves that $(A \cup B) \cap (A \cup C) \subseteq A \cup (B \cap C)$. Hence

$$A \cup (B \cap C) = (A \cup B) \cap (A \cup C).$$

Let $x \in A \cap (B \cup C)$. Then $x \in A$ and $x \in B \cup C$. So $x \in A$, and $x \in B$ or $x \in C$. Thus $x \in A$ and $x \in B$, or $x \in A$ and $x \in C$. Hence $x \in A \cap B$ or $x \in A \cap C$. Therefore $x \in (A \cap B) \cup (A \cap C)$. This argument shows that

$$A \cap (B \cup C) \subseteq (A \cap B) \cup (A \cap C).$$

Reversing the steps above proves that $(A \cap B) \cup (A \cap C) \subseteq A \cap (B \cup C)$. Thus

$$A \cap (B \cup C) = (A \cap B) \cup (A \cap C).$$

This completes the proof of (c).

Let $x \in A - B$. Then $x \in A$ and $x \notin B$. Thus $x \in A$ and $x \in \overline{B}$; therefore $x \in A \cap \overline{B}$. This argument proves that $A - B \subseteq A \cap \overline{B}$. Reversing the steps above proves that $A \cap \overline{B} \subseteq A - B$. Therefore $A - B = A \cap \overline{B}$, proving (i).

31. We have $\overline{(A \cap B)} = \overline{(\overline{\overline{A}} \cap \overline{\overline{B}})} = \overline{\overline{(\overline{A} \cup \overline{B})}} = \overline{A} \cup \overline{B}$ by Theorem 2.1(d).

33. Let $x \in (A \cup B) - (A \cap B)$. Then $x \in A \cup B$ and $x \notin A \cap B$. Therefore $x \in A$ or $x \in B$, and $x \notin A$ or $x \notin B$. So $x \in A$ and $x \notin B$, or $x \in B$ and $x \notin A$. Thus $x \in A - B$ or $x \in B - A$. Hence $x \in (A - B) \cup (B - A)$. This argument proves that

$$(A \cup B) - (A \cap B) \subseteq (A - B) \cup (B - A).$$

Reversing the steps above proves that $(A - B) \cup (B - A) \subseteq (A \cup B) - (A \cap B)$. Therefore

$$(A \cup B) - (A \cap B) = (A - B) \cup (B - A).$$

35. Let $x \in (A - B) - C$. Then $x \in A - B$ and $x \notin C$. So $x \in A$, $x \notin B$, and $x \notin C$. Thus $x \in A$ and $x \notin C$, and $x \notin B$ and $x \notin C$. It follows that $x \in A - C$ and $x \notin B - C$. Hence $x \in (A - C) - (B - C)$, and so $(A - B) - C \subseteq (A - C) - (B - C)$. Reversing the steps proves that $(A - C) - (B - C) \subseteq (A - B) - C$. Thus $(A - B) - C = (A - C) - (B - C)$.

13

37. Let $x \in (A - B) \cap (A - C)$. Then $x \in A - B$ and $x \in A - C$. Therefore $x \in A$ and $x \notin B$, and $x \in A$ and $x \notin C$. Thus $x \in A$, $x \notin B$, and $x \notin C$. So $x \in A$ and $x \notin B \cup C$. Hence $x \in A - (B \cup C)$. Therefore $(A - B) \cap (A - C) \subseteq A - (B \cup C)$. Reversing the steps yields $A - (B \cup C) \subseteq (A - B) \cap (A - C)$. Thus

$$(A - B) \cap (A - C) = A - (B \cup C).$$

39. Let $A = \{1\}$, $B = \{2\}$, $C = \{3\}$, and $D = \{4\}$. Then $(A \times C) \cup (B \times D) = \{(1,3)(2,4)\}$, but $(A \cup B) \times (C \cup D) = \{(1,3), (1,4), (2,3), (2,4)\}$.

2.2 EQUIVALENCE RELATIONS

1. The relation is symmetric because $x \mathrel{R} y$ implies $y \mathrel{R} x$, and it is transitive since $x \mathrel{R} y$ and $y \mathrel{R} z$ implies $x \mathrel{R} z$. It is not reflexive, however, because $3 \mathrel{R} 3$ is false.

3. The relation is reflexive, symmetric, and transitive.

5. The relation is reflexive and symmetric. It is not transitive, however, because if x, y, and z are 66, 67, and 68 inches tall, respectively, then $x \mathrel{R} y$ and $y \mathrel{R} z$ are true but $x \mathrel{R} z$ is false.

7. The relation is reflexive, symmetric, and transitive.

9. The relation is reflexive, symmetric, and transitive.

11. The relation is reflexive and transitive, but it is not symmetric because $\{1\} \mathrel{R} \{1, 2\}$ is true, whereas $\{1, 2\} \mathrel{R} \{1\}$ is false.

13. (i) Since $x - x = 0$ and 0 is even, $x \mathrel{R} x$ holds for every integer x.

(ii) If $x \mathrel{R} y$, then $x - y$ is even. Hence $y - x = -(x - y)$ is even. Therefore $y \mathrel{R} x$.

(iii) If $x \mathrel{R} y$ and $y \mathrel{R} z$, then both $x - y$ and $y - z$ are even. Since the sum of even integers is even, it follows that $x - z = (x - y) + (y - z)$ is even. Hence $x \mathrel{R} z$.

The equivalence class of R containing 7 is the set of odd integers. There are two equivalence classes of R, namely, the sets of even and odd integers.

15. (i) For any integer x greater than 1, the largest prime divisor of x equals the largest prime divisor of x. Thus $x \mathrel{R} x$ holds.

(ii) If the largest prime divisor of x equals the largest prime divisor of y, then obviously the largest prime divisor of y equals the largest prime divisor of x. Hence $x \mathrel{R} y$ implies $y \mathrel{R} x$.

(iii) If the largest prime divisor of x equals the largest prime divisor of y and the largest prime divisor of y equals the largest prime divisor of z, then obviously the largest prime divisor of x equals the largest prime divisor of z. Hence $x\ R\ y$ and $y\ R\ z$ implies $x\ R\ z$.

The largest prime divisor of 60 is 5, and so the equivalence class of 60 contains all the integers greater than 1 that are divisible by 5 but by no prime greater than 5. There are infinitely many distinct equivalence classes of R. For every prime p, there is an equivalence class of R containing the integers greater than 1 that are divisible by p but not by any prime greater than p.

17.   (i) For any ordered pair (x_1, x_2), we have $x_1^2 + x_2^2 = x_1^2 + x_2^2$. Hence $(x_1, x_2)\ R\ (x_1, x_2)$.

  (ii) If $(x_1, x_2)\ R\ (y_1, y_2)$, then $x_1^2 + x_2^2 = y_1^2 + y_2^2$. So $y_1^2 + y_2^2 = x_1^2 + x_2^2$, and therefore $(y_1, y_2)\ R\ (x_1, x_2)$.

  (iii) If $(x_1, x_2)\ R\ (y_1, y_2)$ and $(y_1, y_2)\ R\ (z_1, z_2)$, then

$$x_1^2 + x_2^2 = y_1^2 + y_2^2 \qquad \text{and} \qquad y_1^2 + y_2^2 = z_1^2 + z_2^2.$$

  Thus $x_1^2 + x_2^2 = z_1^2 + z_2^2$, and so $(x_1, x_2)\ R\ (z_1, z_2)$.

For $z = (-3, 4)$, we have $3^2 + (-4)^2 = 25$. Thus the equivalence class of R containing z is the set $\{(x, y)\colon x^2 + y^2 = 25\}$. Geometrically, this set consists of all the points in S that lie on a circle with radius 5 centered at $(0, 0)$. There are infinitely many distinct equivalence classes of R, namely, the sets $\{(x, y)\colon x^2 + y^2 = r\}$ for every nonnegative real number r.

19. The equivalence relation consists of the elements (x, y) such that x and y lie in the same partitioning subset. Hence this equivalence relation is

$$\{(1, 1), (1, 5), (5, 1), (5, 5), (2, 2), (2, 4), (4, 2), (4, 4), (3, 3)\}.$$

21. If $[x] \cap [y] = \varnothing$, it follows from Theorem 2.3(a) that x is not related to y. Thus $x\ R\ y$ is false. On the other hand, If $[x] \cap [y] \neq \varnothing$, then $[x] = [y]$ by Theorem 2.3(b). Hence $x\ R\ y$ by Theorem 2.3(a). Thus by the law of the contrapositive, if $x\ R\ y$ is false, then $[x] \cap [y] = \varnothing$.

23. There may be no element related to x; that is, $x\ R\ y$ may not be true for any y.

25. If S has n elements, then $S \times S$ has n^2 elements. Since a relation on S is a subset of $S \times S$, there are 2^{n^2} relations on S by Theorem 1.3.

27. A partition $\{B, C\}$ of a set S into two subsets consists of a subset A and its complement $S - A$, both of which must be nonempty. Thus if S contains n elements, there are $2^n - 2$ possible choices for B, namely any subset of S except $\varnothing$ and S. However this number counts every partition twice, once with $A = B$ and once with $A = C$. Thus there are only

$$\frac{1}{2}(2^n - 2) = 2^{n-1} - 1$$

distinct partitions of S into two subsets.

29. There are 15 partitions of a set with four elements: 1 consisting of a single subset, 7 consisting of two subsets, 6 consisting of three subsets, and 1 consisting of four subsets.

31. (i) Because $f(s) = f(s)$ for each $s \in S$, R is reflexive.

(ii) If $x \ R \ y$, then $f(x) = f(y)$. But then $f(y) = f(x)$; so $y \ R \ x$ and R is symmetric.

(iii) If $x \ R \ y$ and $y \ R \ z$, then $f(x) = f(y)$ and $f(y) = f(z)$. Hence $f(x) = f(z)$, and so R is transitive.

33. Let $A = \{1, 2, \ldots, n, n+1\}$. The number of partitions of A into m subsets is $p_m(n+1)$. In any such partition, either $n + 1$ occurs alone in some partitioning subset or $n + 1$ is accompanied by another element of A.

Type 1: $n + 1$ occurs alone.
Since $\{n + 1\}$ is one of the partitioning subsets, the other $m - 1$ partitioning subsets form a partition of $\{1, 2, \ldots, n\}$. Thus there are $p_{m-1}(n)$ such partitions.

Type 2: $n + 1$ is accompanied by another element of A.
If we delete $n + 1$ from its partitioning subset, then the resulting sets form a partition of $\{1, 2, \ldots, n\}$ into m subsets. Since there are m different subsets from which $n + 1$ could have been removed, the multiplication principle shows that there are $m p_m(n)$ such partitions.

Because every partition of A into m subsets is either of type 1 or type 2 above, the number of such partitions is $p_{m-1}(n) + m p_m(n)$. Hence

$$p_m(n + 1) = p_{m-1}(n) + m p_m(n).$$

2.3 PARTIAL ORDERING RELATIONS

1. Since $(2, 2) \notin R$, relation R is not reflexive. Hence R is not a partial order on S.

3. Since R is reflexive, antisymmetric, and transitive on S, R is a partial order on S.

5. Since R is reflexive, antisymmetric, and transitive on S, R is a partial order on S.

7. We have $[2] \ R \ [3]$ and $[3] \ R \ [4]$, but $[2] \ R \ [4]$ is false. Hence R is not transitive, and so R is not a partial order on S.

9. In a Hasse diagram for R, each element of S is represented by a point, and a segment is drawn upward from the point representing x to the point representing y whenever $x \ R \ y$ and there is no element $s \in S$ such that both $x \ R \ s$ and $s \ R \ y$ are true. Thus the Hasse diagram for R is as follows.

11. Proceeding as in Exercise 9, we obtain the following Hasse diagram.

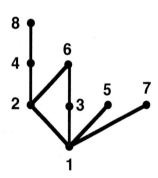

13. The set of elements for this relation is $S = \{2, 3, 6, 12\}$. Each element of S is related to itself and to any element in the Hasse diagram that can be reached by moving upward along edges in the diagram. Thus 2 is related to 2, 6, and 12; 3 is related to 3, 6, and 12; 6 is related to 6 and 12; and 12 is related to itself. Thus the relation is

$$\{(2,2), (2,6), (2,12), (3,3), (3,6), (3,12), (6,6), (6,12), (12,12)\}.$$

15. As in Exercise 13, we find that the relation is $\{(2,2), (x,x), (x,A), (A,A), (\varnothing, \varnothing)\}$.

17. An element $x \in S$ is a minimal element of S if the only element of S that divides x is x itself. Since 1 divides each element of S, it follows that 1 is the only minimal element of S. Likewise, an element $x \in S$ is a maximal element of S, if x divides no element of S except x itself. Thus the elements 4, 5, and 6 in S are maximal elements of S.

19. As in Exercise 17, we see that 2, 3, and 4 are minimal elements in S, and 1 and 2 are maximal elements in S.

21. A Hasse diagram for R is shown below.

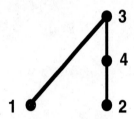

We will apply the topological sorting algorithm to this Hasse diagram. First, we set $S' = S$. The minimal elements of S' are 1 and 2. We select one of these arbitrarily, say $s_1 = 1$, and delete 1 from S', so that $S' = \{2, 3, 4\}$. Second, we choose a minimal element of S'. The only minimal element is 2, and so we set $s_2 = 2$. Deleting 2 from S' produces $S' = \{3, 4\}$. Third, we again pick a minimal element of S'. The only choice is $s_3 = 4$, and we obtain $S' = \{3\}$ after deleting 4 from S'. Fourth, we pick the only remaining element of S', which is $s_4 = 3$. Thus one sequence in which the elements of S can be chosen is 1, 2, 4, 3.

23. The elements of S are $\varnothing, \{1\}, \{2\}, \{3\}, \{1, 2\}$, and $\{1, 3\}$. A Hasse diagram for the relation R is shown below.

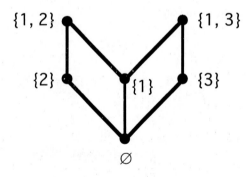

Using the topological sorting algorithm as in Exercise 21, we select the elements $A_1 = \varnothing$, $A_2 = \{1\}$, $A_3 = \{2\}$, $A_4 = \{3\}$, $A_5 = \{1, 2\}$, and $A_6 = \{1, 3\}$. Hence one sequence in which the elements of S can be chosen is

$$\varnothing, \{1\}, \{2\}, \{3\}, \{1, 2\}, \{1, 3\}.$$

25. The relation is not antisymmetric because $2 \ R \ (-2)$ and $(-2) \ R \ 2$, but $2 \neq -2$.

27. For the "divides" relation to be a total order on S, one element in each pair of elements of S must divide the other. One such set S is $\{1, 2, 4, 8, 16\}$.

29. Consider the set $S = \{2, 3, 4, 5, 6, 9, 15\}$ and the relation R defined by $x \, R \, y$ if and only if x divides y. A minimal element of S is one that is divisible by no element of S except itself. Therefore, 2, 3 and 5 are the minimal elements of S. A maximal element of S is one that cannot be divided into any element of S except itself. Thus 4, 6, 9, and 15 are the maximal elements of S. Hence S has exactly three minimal elements and exactly four maximal elements.

31. Let R be a total order on a set S with five elements. By Theorem 2.8, S must have a minimal element with respect to R. Because every pair of elements in S are related, it follows that this minimal element a is unique. Delete a from S to obtain a set S_1, which, as above, must have a unique minimal element b. By continuing in this manner, we see that the Hasse diagram for S must be as follows.

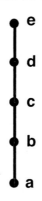

33. Let $x \in S'$. Because R is reflexive, $(x, x) \in R$. Thus $(x, x) \in R \cap (S' \times S') = R'$. Hence R' is reflexive. If $x, y \in S'$, $(x, y) \in R'$, and $(y, x) \in R'$, then $(x, y) \in R$ and $(y, x) \in R$. Therefore $x = y$ since R is antisymmetric, and so R' is antisymmetric. Finally if $x, y, z \in S'$, $(x, y) \in R'$, and $(y, z) \in R'$, then $(x, y) \in R$ and $(y, z) \in R$. Hence $(x, z) \in R$ because R is transitive. Thus $(x, z) \in R \cap (S' \times S') = R'$, and so R' is transitive. Because R' is reflexive, antisymmetric, and transitive, R' is a partial order on S'.

35. Let R be a total order on S, and let x and y be minimal elements in S. Because R is a total order on S, we must have $x \, R \, y$ or $y \, R \, x$. If $x \, R \, y$, then we must have $x = y$ because y is a minimal element of S. Likewise $y \, R \, x$ implies $y = x$. Since in either case $x = y$, it follows that a minimal element of S is unique.

37. Let R_1 be a total order on S_1 and R_2 be a total order on S_2. Let R denote the lexicographic order on $S_1 \times S_2$. Recall that R is a partial order by Theorem 2.7. Given (u, v) and (x, y) in $S_1 \times S_2$, the elements u and x must be comparable in S_1 and v and y must be comparable in S_2. If $u \, R_1 \, x$ and $u \neq x$, then $(u, v) \, R \, (x, y)$ by the definition of R. Likewise, if $x \, R_1 \, u$ and $u \neq x$, then $(x, y) \, R \, (u, v)$. Finally, if $u = x$, then $(u, v) \, R \, (x, y)$ if $v \, R \, y$, and $(x, y) \, R \, (u, v)$ if $y \, R \, v$. Thus in any of the cases (u, v) and (x, y) can be compared. Hence R is a total order on $S_1 \times S_2$.

39. **(a)** Let R_1 be a partial order on S_1, R_2 be a partial order on S_2, and R denote the lexicographic order on $S_1 \times S_2$. Let a be a minimal element of S_1 and b be a minimal element of S_2. Suppose that $(u,v)\ R\ (a,b)$. Then $u\ R_1\ a$. Since a is a minimal element of S_1, it follows that $u = a$. Now $(u,v)\ R\ (a,b)$ and $u = a$ implies that $v\ R_2\ b$. Because b is a minimal element of S_2 and $v\ R_2\ b$, we have $v = b$. Hence $(u,v) = (a,b)$, proving that (a,b) is a minimal element of $S_1 \times S_2$ with respect to the lexicographic order.

(b) Let R_1 be a partial order on S_1, R_2 be a partial order on S_2, and R denote the lexicographic order on $S_1 \times S_2$. Let a be a maximal element of S_1 and b be a maximal element of S_2. We will prove that (a,b) is a maximal element of $S_1 \times S_2$. Suppose that $(a,b)\ R\ (u,v)$. Then $a\ R_1\ u$. Since a is a maximal element of S_1, it follows that $u = a$. Now $(a,b)\ R\ (u,v)$ and $u = a$ imply that $b\ R_2\ v$. Because b is a maximal element of S_2 and $b\ R_2\ v$, we have $v = b$. Hence $(u,v) = (a,b)$, proving that (a,b) is a maximal element of $S_1 \times S_2$ with respect to the lexicographic order.

41. Let S be a set containing exactly n elements, and let T be a total order on S. By Theorem 2.8, there is a minimal element s_1 in S. Moreover, s_1 is unique by Exercise 35. Let S_1 be the set obtained by deleting s_1 from S. Then, as in Exercise 33, it can be shown that $T_1 = T \cap (S_1 \times S_1)$ is a total order on S_1. Hence S_1 contains a unique minimal element s_2. Continuing in this manner, we obtain a sequence of elements $s_1, s_2, \ldots, s_n$, where s_{k+1} is the unique minimal element of S_k, the set obtained by removing s_k from S_{k-1}. Conversely, every sequence $a_1, a_2, \ldots, a_n$ of the elements in S gives rise to a unique total order R on S by defining $(a_i, a_j) \in R$ if and only if $i \leq j$. Thus the number of total orders on S is the same as the number of permutations of the elements of S, which is $n!$.

2.4 FUNCTIONS

1. Since every element of X is the first entry of a unique ordered pair in R, R is a function with domain X.

3. The relation R is not a function with domain X because there is no ordered pair having -1 as its first entry.

5. The relation g is a function with domain X.

7. The relation g is not a function with domain X because some students have no older brother.

9. The relation g is not a function with domain X because the logarithmic function with base 2 is defined only for positive real numbers.

11. The relation g is a function with domain X.

13. $f(3) = 5(3) - 7 = 8$

15. $f(-2) = 2^{-2} = \dfrac{1}{2^2} = \dfrac{1}{4}$

17. $f(9) = \sqrt{9 - 5} = \sqrt{4} = 2$

19. $f(-3) = -(-3)^2 = -9$

21. $\log_2 8 = \log_2 (2^3) = 3$

23. $\log_2 1 = \log_2 (2^0) = 0$

25. $\log_2 \dfrac{1}{16} = \log_2 \dfrac{1}{2^4} = \log_2 (2^{-4}) = -4$

27. $\log_2 \dfrac{1}{32} = \log_2 \dfrac{1}{2^5} = \log_2 (2^{-5}) = -5$

29. $\log_2 37 = \dfrac{\log 37}{\log 2} \approx 5.21$

31. $\log_2 0.86 = \dfrac{\log 0.86}{\log 2} \approx -0.22$

33. $\log_2 1.54 = \dfrac{\log 1.54}{\log 2} \approx 0.62$

35. $\log_2 1000 = \dfrac{\log 1000}{\log 2} \approx 9.97$

37. $gf(x) = g(f(x)) = 2(4x + 7) - 3 = 8x + 11; \, fg(x) = f(g(x)) = 4(2x - 3) + 7 = 8x - 5$

39. $gf(x) = 5(2^x) + 7; \, fg(x) = 2^{5x+7}$

41. $gf(x) = |x|(\log_2 |x|); \, fg(x) = |x \log_2 x|$

43. $gf(x) = x^2 - 2x + 1; \, fg(x) = (x + 1)^2 - 2(x + 1) = x^2 - 1$

45. This function is one-to-one because $f(x_1) = f(x_2)$ if and only if $3x_1 = 3x_2$ if and only if $x_1 = x_2$. It is not onto since its range is the set of multiples of three.

47. This function is one-to-one because $f(x_1) = f(x_2)$ if and only if $3 - x_1 = 3 - x_2$ if and only if $x_1 = x_2$. It is also onto because, for any $y \in Z$, $3 - y \in Z$ and $g(3 - y) = y$.

49. This function is onto because, for any $y \in Z$, $g(2y) = y$. It is not one-to-one because $g(1) = g(2)$.

51. This function is neither one-to-one nor onto since $g(1) = g(-1)$ and the range of g is the set of nonnegative integers.

53. We proceed as in Example 2.47. Since $y = 5x$ implies $\frac{y}{5} = x$, $f^{-1}(x) = \frac{x}{5}$.

55. Since $y = -x$ implies $-y = x$, $f^{-1}(x) = -x$.

57. Since $y = \sqrt[3]{x}$ implies $y^3 = x$, $f^{-1}(x) = x^3$.

59. Since $f(x) > 0$ for all real numbers x, the function f is not onto and hence has no inverse.

61. Take Y to be the set of positive real numbers. Then g is a one-to-one correspondence. Moreover,

$$y = 3 \cdot 2^{x+1}$$

$$\frac{y}{3} = 2^{x+1}$$

$$\log_2 \frac{y}{3} = x + 1$$

$$-1 + \log_2 \frac{y}{3} = x.$$

Thus

$$g^{-1}(x) = -1 + \log_2 \frac{x}{3}.$$

63. Suppose that a function f with domain X and codomain Y is to be constructed. Then for each element $x \in X$ there are n possible choices for $f(x)$, namely each of the elements of Y. Thus, by the multiplication principle, there are n^m choices to be made in defining f. Consequently there are n^m different functions with domain X and codomain Y.

65. Let $f\colon X \to Y$ and $g\colon Y \to Z$ be one-to-one functions, and suppose that $gf(x_1) = gf(x_2)$. Then $g(f(x_1) = g(f(x_2))$. Since g is one-to-one, it follows that $f(x_1) = f(x_2)$. But then $x_1 = x_2$ because f is one-to-one. Thus $gf(x_1) = gf(x_2)$ implies $x_1 = x_2$, and hence gf is one-to-one.

67. Let $z \in Z$. Because gf is onto, there exists an $x \in X$ with $z = gf(x)$. Let $y = f(x)$. Then $y \in Y$, and $g(y) = g(f(x)) = gf(x) = z$. Hence g is onto.

To see that f need not be onto, let $X = \{1, 2\}$, $Y = \{3, 4\}$, and $Z = \{5\}$. Define $f\colon X \to Y$ by $f(1) = f(2) = 3$, and define $g\colon Y \to Z$ by $g(3) = g(4) = 5$. Then gf is onto, but f is not.

69. If both f and g are one-to-one correspondences, then gf is a one-to-one correspondence by Exercises 65 and 66. In this case $(gf)^{-1}$ satisfies $(gf)^{-1}(z) = x$ if and only if $gf(x) = z$. But if $gf(x) = z$, then $g^{-1}(z) = f(x)$, and so $f^{-1}(g^{-1}(z)) = x$. Since this equality holds for all $z \in Z$, we have $(gf)^{-1} = f^{-1}g^{-1}$.

2.5   MATHEMATICAL INDUCTION

1. The first two Fibonacci numbers are 1, and each subsequent number is the sum of the two preceding numbers. Thus the first ten Fibonacci numbers are:

$$1, 1, 2, 3, 5, 8, 13, 21, 34, 55.$$

3. We are given that $x_1 = 3$ and $x_2 = 4$. Then

$$
\begin{aligned}
x_3 &= x_2 + x_1 = 4 + 3 = 7, \\
x_4 &= x_3 + x_2 = 7 + 4 = 11, \\
x_5 &= x_4 + x_3 = 11 + 7 = 18, \\
x_6 &= x_5 + x_4 = 18 + 11 = 29, \\
x_7 &= x_6 + x_5 = 29 + 18 = 47, \\
x_8 &= x_7 + x_6 = 47 + 29 = 76.
\end{aligned}
$$

5. Let x_n denote the nth even positive integer. Since the first even positive integer is 2 and each subsequent even positive integer is two more than its predecessor, we can define the even positive integers as follows:

$$
x_n = \begin{cases} 2 & \text{if } n = 1 \\ x_{n-1} + 2 & \text{if } n \geq 2. \end{cases}
$$

7. No base for the induction was established.

9. The proof of the inductive step is faulty because $x - 1$ and $y - 1$ need not be *positive* integers. (If $x = 1$, then $x - 1 = 0$; and if $y = 1$, then $y - 1 = 0$.) Hence the induction hypothesis cannot be applied to $x - 1$ and $y - 1$.

11. If $n = 1$, the left side of the equation to be proved is 1, and the right side is

$$\frac{1(1 + 1)}{2} = 1.$$

Hence the equation to be proved is true if $n = 1$. Assume that the equation holds for $n = k \geq 1$, that is,

$$1 + 2 + \cdots + k = \frac{k(k + 1)}{2}.$$

23

Then

$$1 + 2 + \cdots + k + (k + 1) = (1 + 2 + \cdots + k) + (k + 1)$$

$$= \frac{k(k + 1)}{2} + (k + 1)$$

$$= \frac{k(k + 1) + 2(k + 1)}{2}$$

$$= \frac{(k + 2)(k + 1)}{2}$$

$$= \frac{(k + 1)(k + 2)}{2}.$$

This proves the equation for $n = k + 1$. Hence, by the principle of mathematical induction, the equation is true for all positive integers n.

13. If $n = 1$, the left side of the equation to be proved is 1, and the right side is

$$\frac{1^2(1 + 1)^2}{4} = \frac{2^2}{4} = 1.$$

So the equation is true for $n = 1$. Assume that the equation holds for $n = k \geq 1$. Then

$$1 + 8 + 27 + \cdots + k^3 = \frac{k^2(k + 1)^2}{4}.$$

Hence

$$1 + 8 + 27 + \cdots + k^3 + (k + 1)^3 = (1 + 8 + 27 + \cdots + k^3) + (k + 1)^3$$

$$= \frac{k^2(k + 1)^2}{4} + (k + 1)^3$$

$$= \frac{k^2(k + 1)^2 + 4(k + 1)^3}{4}$$

$$= \frac{\left[k^2 + 4(k + 1)\right](k + 1)^2}{4}$$

$$= \frac{\left(k^2 + 4k + 4\right)(k + 1)^2}{4}$$

$$= \frac{(k + 2)^2(k + 1)^2}{4}.$$

Thus the equation is proved for $n = k + 1$. Hence, by the principle of mathematical induction, the equation is true for all positive integers n.

15. If $n = 1$, both sides of the equation to be proved equal 1. So the equation is true for $n = 1$. Assume that the equation holds for $n = k \geq 1$. Then

$$1(1!) + 2(2!) + \cdots + k(k!) + (k+1)(k+1)! = [(k+1)! - 1] + (k+1)(k+1)!$$
$$= (k+1)![1 + (k+1)] - 1$$
$$= (k+1)!(k+2) - 1$$
$$= (k+2)! - 1.$$

This proves the equation for $n = k + 1$. Hence, by the principle of mathematical induction, the equation is true for all positive integers n.

17. For $n = 2$, the left side is $1 \cdot 3 = 3$ and the right side is 2; so the inequality holds for $n = 2$. Assume that it holds for $n = k \geq 2$; that is,

$$1 \cdot 3 \cdots (2k - 1) \geq 2 \cdot 4 \cdots (2k - 2).$$

Since all the factors on both sides of this inequality are positive, we have

$$1 \cdot 3 \cdots (2k - 1) \cdot (2k + 1) \geq 2 \cdot 4 \cdots (2k - 2) \cdot (2k + 1)$$
$$\geq 2 \cdot 4 \cdots (2k - 2) \cdot 2k,$$

which proves the inequality for $n = k + 1$. Hence, by the principle of mathematical induction, the inequality is true for all integers $n \geq 2$.

19. Since $7! = 5040$ and $3^7 = 2187$, $n! > 3n$ if $n = 7$. Assume that $k! > 3k$ for some integer $k \geq 7$. Then

$$(k + 1)! = (k + 1)k! > (k + 1)3^k > 3(3^k) = 3^{k+1}.$$

Thus the inequality holds for $n = k + 1$. Hence, by the principle of mathematical induction, the inequality holds for all integers $n \geq 7$.

21. Since $F_2 = 1 \leq 2 = 2F_1$ and $F_3 = 2 = 2F_2$, the inequality holds if $n = 2$ or $n = 3$. Assume that for some integer $k \geq 3$, the inequality holds for all integers n such that $2 \leq n \leq k$. Then

$$F_{k+1} = F_k + F_{k-1} \leq 2F_{k-1} + 2F_{k-2} = 2(F_{k-1} + F_{k-2}) = 2F_k.$$

Thus the inequality holds for $n = k+1$. Hence, by the strong principle of mathematical induction, the inequality holds for all integers $n \geq 2$.

23. Since $F_2 = 1 = 2 - 1 = F_3 - 1$, the equation holds if $n = 1$. Assume that the equation holds for $n = k \geq 1$. Then

$$F_2 + F_4 + \cdots + F_{2k} + F_{2(k+1)} = (F_{2k+1} - 1) + F_{2k+2}$$
$$= (F_{2k+1} + F_{2k+2}) - 1$$
$$= F_{2k+3} - 1.$$

Thus the equation holds for $n = k + 1$. Hence, by the principle of mathematical induction, the equation holds for all positive integers n.

25. If $n = 3$, then

$$F_n = 2 \geq \frac{125}{64} = \left(\frac{5}{4}\right)^3.$$

Likewise, if $n = 4$, then

$$F_n = 3 \geq \frac{625}{256} = \left(\frac{5}{4}\right)^4.$$

So the inequality holds for $n = 3$ and $n = 4$. Assume that the inequality holds for $n = 3, 4, \ldots, k$. Then

$$F_{k+1} = F_k + F_{k-1}$$
$$\geq \left(\frac{5}{4}\right)^k + \left(\frac{5}{4}\right)^{k-1}$$
$$= \left(\frac{5}{4}\right)^{k-1}\left(\frac{5}{4} + 1\right)$$
$$= \frac{9}{4} \cdot \left(\frac{5}{4}\right)^{k-1}$$
$$\geq \frac{25}{16} \cdot \left(\frac{5}{4}\right)^{k-1}$$
$$= \left(\frac{5}{4}\right)^{k+1}$$

Thus the inequality holds for $n = k + 1$. Hence, by the principle of mathematical induction, the inequality holds for all integers $n \geq 3$.

27. If $r = 1$, then $s_0 + s_1 + \cdots + s_n = s_0 + s_0 + \cdots + s_0 = s_0(n + 1)$. If $r \neq 1$, then by the equation in Example 2.51, we have

$$s_0 + s_1 + \cdots + s_n = s_0 r^0 + s_0 r^1 + \cdots + s_0 r^n$$
$$= s_0(r^0 + r^1 + \cdots + r^n)$$
$$= s_0(1 + r + \cdots + r^n)$$
$$= \frac{s_0(r^{n+1} - 1)}{r - 1}.$$

We will prove this formula by induction on n. If $n = 0$, then the left side is s_0 and the right side is

$$\frac{s_0(r^{n+1} - 1)}{r - 1} = \frac{s_0(r^1 - 1)}{r - 1} = s_0.$$

Thus the formula is true for $n = 0$. Assume that the formula holds for some nonnegative integer k. Then

$$s_0 + s_1 + \cdots + s_k + s_{k+1} = (s_0 + s_1 + \cdots + s_k) + s_{k+1}$$

$$= \frac{s_0(r^{k+1} - 1)}{r - 1} + s_0 r^{k+1}$$

$$= \frac{s_0(r^{k+1} - 1) + s_0 r^{k+1}(r - 1)}{r - 1}$$

$$= \frac{s_0(r^{k+2} - 1)}{r - 1},$$

which proves the formula for $n = k + 1$. Hence, by the principle of mathematical induction, the formula holds for all nonnegative integers n.

2.6 APPLICATIONS

1. $C(7, 2) = \dfrac{7!}{2!\,(7-2)!} = \dfrac{7 \cdot 6 \cdot 5!}{2!\,5!} = \dfrac{7 \cdot 6}{2 \cdot 1} = \dfrac{42}{2} = 21$

3. $C(10, 5) = \dfrac{10!}{5!\,(10-5)!} = \dfrac{10 \cdot 9 \cdot 8 \cdot 7 \cdot 6 \cdot 5!}{5!\,5!} = \dfrac{10 \cdot 9 \cdot 8 \cdot 7 \cdot 6}{5 \cdot 4 \cdot 3 \cdot 2 \cdot 1} = \dfrac{30{,}240}{120} = 252$

5. $C(11, 4) = \dfrac{11!}{4!\,(11-4)!} = \dfrac{11 \cdot 10 \cdot 9 \cdot 8 \cdot 7!}{4!\,7!} = \dfrac{11 \cdot 10 \cdot 9 \cdot 8}{4 \cdot 3 \cdot 2 \cdot 1} = \dfrac{7920}{24} = 330$

7. $C(11, 6) = \dfrac{11!}{6!\,(11-6)!} = \dfrac{11 \cdot 10 \cdot 9 \cdot 8 \cdot 7 \cdot 6!}{6!\,5!} = \dfrac{11 \cdot 10 \cdot 9 \cdot 8 \cdot 7}{5 \cdot 4 \cdot 3 \cdot 2 \cdot 1} = \dfrac{55{,}440}{120} = 462$

9. $C(n, 0) = \dfrac{n!}{0!\,n!} = \dfrac{n!}{1 \cdot n!} = 1$

11. $C(n, 2) = \dfrac{n!}{2!\,(n-2)!} = \dfrac{n(n-1)(n-2)!}{2!\,(n-2)!} = \dfrac{n(n-1)}{2}$

13. By Theorem 1.3, there are $2^6 = 64$ subsets.

15. A pizza can be ordered with any subset of the set of 7 toppings. Hence there are $2^7 = 128$ possible pizzas that can be ordered.

17. As in Exercise 15, the car can be equipped in $2^8 = 256$ different ways.

19. By Theorem 2.10, the number of 5-element subsets is $C(7, 5) = 21$.

21. There are $C(12, 5) = 792$ possible starting teams.

23. There are $C(6,3) = 20$ choices of three vegetables.

25. There are $C(10,3) = 120$ sets of winning candidates.

27. There are $C(52,13) = \dfrac{52!}{13!\,39!}$ possible bridge hands.

29. Let x and y be distinct real numbers. For $n = 0$, the left side of the equation is $x^0 y^0 = 1$ and the right side is

$$\frac{x^1 - y^1}{x - y} = 1.$$

Thus the equation is true for $n = 0$. Assume that the equation is true for some nonnegative integer k. Then

$$x^{k+1}y^0 + x^k y^1 + \cdots + x^1 y^k + x^0 y^{k+1} = x\left(x^k y^0 + x^{k-1}y^1 + \cdots + x^1 y^{k-1} + x^0 y^k\right) + y^{k+1}$$

$$= x\left(\frac{x^{k+1} - y^{k+1}}{x - y}\right) + y^{k+1}$$

$$= \frac{x\left(x^{k+1} - y^{k+1}\right) + y^{k+1}(x - y)}{x - y}$$

$$= \frac{x^{k+2} - y^{k+2}}{x - y},$$

so that the equation is true for $n = k+1$. Therefore, by the principle of mathematical induction, the equation holds for all nonnegative integers n.

31. If $n = 1$, then

$$\frac{(2n)!}{2^n} = \frac{2}{2} = 1$$

is an integer. Assume that $(2k)!/2^k$ is an integer m. Then

$$\frac{[2(k+1)]!}{2^{k+1}} = \frac{(2k+2)!}{2^{k+1}} = \frac{(2k+2)(2k+1)(2k)!}{2 \cdot 2^k}$$

$$= \frac{(2k+2)(2k+1)m}{2} = (k+1)(2k+1)m$$

is an integer. Therefore, for all positive integers n, $(2n)!/2^n$ is an integer.

33. If $n = 1$, then 3 divides $2^{2n} - 1 = 2^2 - 1 = 3$. Assume that 3 divides $2^{2k} - 1$ for some positive integer k, and let $2^{2k} - 1 = 3m$. Then

$$2^{2(k+1)} - 1 = 2^{2k+2} - 1 = 4 \cdot 2^{2k} - 1 = 3 \cdot 2^{2k} + 2^{2k} - 1 = 3(2^{2k} + m).$$

Hence 3 divides $2^{2(k+1)} - 1$. Thus, by the principle of mathematical induction, 3 divides $2^{2n} - 1$ for all positive integers n.

35. If $n = 0$, then

$$\frac{(4n)!}{8^n} = \frac{0!}{8^0} = \frac{1}{1} = 1$$

is an integer. Assume that for some nonnegative integer k, $(4k)!/8^k$ is an integer m. Then

$$\frac{[4(k+1)]!}{8^{k+1}} = \frac{(4k+4)!}{8 \cdot 8^k} = \frac{(4k+4)(4k+3)(4k+2)(4k+1)(4k)!}{8 \cdot 8^k}$$

$$= \frac{4(k+1)(4k+3)[2(2k+1)](4k+1)m}{8}$$

$$= (k+1)(4k+3)(2k+1)(4k+1)m.$$

Therefore, for all positive integers n, $(4n)!/8^n$ is an integer.

37. If $n = 1$, then both sides of the equation to be proved equal 1. Assume that

$$(1 + 2 + \cdots + k)^2 = 1^3 + 2^3 + \cdots + k^3$$

for some positive integer k. Then, by Exercise 11 in Section 2.5, we have

$$[1 + 2 + \cdots + k + (k+1)]^2 = (1 + 2 + \cdots + k)^2 + 2(1 + 2 + \cdots + k)(k+1) + (k+1)^2$$

$$= (1 + 2 + \cdots + k)^2 + 2\frac{k(k+1)}{2}(k+1) + (k+1)^2$$

$$= (1^3 + 2^3 + \cdots + k^3) + k(k+1)^2 + (k+1)^2$$

$$= (1^3 + 2^3 + \cdots + k^3) + (k+1)(k+1)^2$$

$$= 1^3 + 2^3 + \cdots + k^3 + (k+1)^3.$$

Thus the equation is proved for $n = k + 1$. Hence, by the principle of mathematical induction, the equation is true for all positive integers n.

39. If $n = 2$, then

$$\frac{1}{\sqrt{1}} + \frac{1}{\sqrt{2}} = 1 + \frac{\sqrt{2}}{2} > \frac{\sqrt{2}}{2} + \frac{\sqrt{2}}{2} = \sqrt{2}.$$

So the inequality holds if $n = 2$. Assume that for some integer $k \geq 2$

$$\frac{1}{\sqrt{1}} + \frac{1}{\sqrt{2}} + \cdots + \frac{1}{\sqrt{k}} > \sqrt{k}.$$

Then

$$\frac{1}{\sqrt{1}} + \frac{1}{\sqrt{2}} + \cdots + \frac{1}{\sqrt{k}} + \frac{1}{\sqrt{k+1}} > \sqrt{k} + \frac{1}{\sqrt{k+1}}$$

$$= \frac{\sqrt{k}\sqrt{k+1}}{\sqrt{k+1}} + \frac{1}{\sqrt{k+1}}$$

$$> \frac{\sqrt{k}\sqrt{k}}{\sqrt{k+1}} + \frac{1}{\sqrt{k+1}}$$

$$= \frac{k+1}{\sqrt{k+1}}$$

$$= \sqrt{k+1}.$$

Hence the inequality holds for $n = k+1$. Therefore it holds for all integers $n \geq 2$ by the principle of mathematical induction.

41. For $n = 2$, the equation to be proved is one of De Morgan's laws established in Theorem 2.2(a). Assume that for some integer $k \geq 2$ we have

$$\overline{(A_1 \cup A_2 \cup \cdots \cup A_k)} = \overline{A_1} \cap \overline{A_2} \cap \cdots \cap \overline{A_k}.$$

Then, by Theorem 2.2(a) and the induction hypothesis, we have

$$\overline{(A_1 \cup A_2 \cup \cdots \cup A_k \cup A_{k+1})} = \overline{A_1 \cup A_2 \cup \cdots \cup A_k} \cap \overline{A_{k+1}}$$
$$= \overline{A_1} \cap \overline{A_2} \cap \cdots \cap \overline{A_k} \cap \overline{A_{k+1}}.$$

Hence the equality holds for $n = k+1$. Therefore the equation holds for all integers $n \geq 2$ by the principle of mathematical induction.

43. There are $n(n-3)/2$ diagonals in a regular n-sided polygon. For $n = 3$, we have

$$\frac{n(n-3)}{2} = \frac{3(3-3)}{2} = \frac{0}{2} = 0,$$

which is the number of diagonals in an equilateral triangle. So the formula is correct if $n = 3$. Assume that the number of diagonals in a regular k-sided polygon is $k(k-3)/2$ for some integer $k \geq 3$. In a regular $(k+1)$-sided polygon, there are $k-2$ diagonals from the additional vertex and one other new diagonal. Hence there are $k-1$ more diagonals than in a regular k-sided polygon. Thus if $n = k+1$, then the number of diagonals in a regular n-sided polygon is

$$\frac{k(k-3)}{2} + (k-1) = \frac{k(k-3) + 2(k-1)}{2} = \frac{k^2 - k - 2}{2} = \frac{(k+1)(k-2)}{2}.$$

Therefore the formula is correct for $n = k+1$, and so it is correct for all integers $n \geq 3$ by the principle of mathematical induction.

45. A list containing $1 = 2^0$ number can be sorted using $0 \cdot 2^0 = 0$ comparisons. Hence the conclusion holds for $n = 0$. Assume that for some nonnegative integer k, a list of 2^k numbers can be sorted into nondecreasing order using at most $k \cdot 2^k$ comparisons. Given a list of 2^{k+1} numbers, divide it into two sublists containing 2^k numbers each. Order each sublist using at most $k \cdot 2^k$ comparisons per sublist. Then merge the two sublists into one ordered list (as in Example 2.56) using at most $2^{k+1} - 1$ comparisons. Then the total number of comparisons required to sort the original list is

$$2k \cdot 2^k + (2^{k+1} - 1) = k \cdot 2^{k+1} + 2^{k+1} - 1 = (k+1)2^{k+1} - 1 < (k+1)2^{k+1}.$$

Hence the conclusion holds for $n = k + 1$. Therefore the conclusion is correct for all nonnegative integers n by the principle of mathematical induction.

47. Mr. and Mrs. Lewis both shook n hands. The result is trivial if $n = 0$ because no hands can be shaken. Assume the conclusion is true if $n = k$, and consider a party with $k + 1$ couples that satisfies the given conditions. Note that no one can shake more than $2k + 2$ hands because there are only $2k + 4$ persons at the party. Since the answers to Mr. Lewis's question are all different, the answers must have been $0, 1, 2, \ldots, 2k + 2$. The persons who shook 0 hands and $2k + 2$ hands are spouses, for otherwise the person shaking $2k + 2$ hands would not have shaken hands with at least three people. Send this couple (which does not include either Mr. or Mrs. Lewis) out of the room. Then the remaining persons satisfy the induction hypothesis, and hence in this smaller group Mr. and Mrs. Lewis have each shaken k hands. But then Mr. and Mrs. Lewis must each have shaken $k + 1$ hands in the original group. Thus the conclusion is true for $n = k+1$. Therefore the conclusion is correct for all nonnegative integers n by the principle of mathematical induction.

SUPPLEMENTARY EXERCISES

1. The intersection of A and C contains all of the elements common to both A and C; so $A \cap C = \{3\}$.

3. The complement of set A contains all of the elements in the universal set U that are not in A; so $\overline{A} = \{5, 6\}$.

5. As in Exercises 3 and 1, we have $\overline{A} \cap \overline{B} = \{6\}$.

7. The union of B and C contains all of the elements in B or C or both; so

$$B \cup C = \{1, 3, 4, 5, 6\}.$$

Hence $\overline{B \cup C} = \{2\}$.

9. A Venn diagram for $A - B$ is shown in Figure 2.2 of the textbook. The desired set is the complement of $A - B$, and so a Venn diagram for it is as shown below.

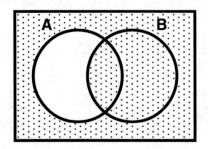

11. First create a Venn diagram for $B - C$ and then one for the union of $\overline{A}$ and $B - C$. The resulting diagram is shown below.

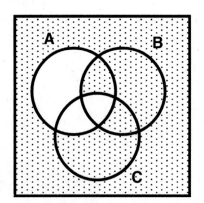

13. $gf(x) = 1 - 4(2x + 3)^2 = 1 - 4(4x^2 + 12x + 9) = -16x^2 - 48x - 35$
    $fg(x) = 2(1 - 4x^2) + 3 = 5 - 8x^2$

15. Since every element of X is the first entry of a unique ordered pair in R, R is a function with domain X.

17. Since every element of X is the first entry of a unique ordered pair in R, R is a function with domain X.

19. The range of g contains only odd integers, and so g is not onto. However, g is one-to-one because if $g(x_1) = g(x_2)$, then

$$2x_1 - 7 = 2x_2 - 7$$
$$2x_1 = 2x_2$$
$$x_1 = x_2.$$

21. Since $g(1) = -1$ and $g(-4) = -1$, g is not one-to-one. However, g is onto, for if $z \in Z$, then $g(z-3) = z$ if $z \leq 0$ and $g(z+2) = z$ if $z > 0$. Thus every codomain element z is the image of some domain element.

23. Because $f(1) = -1$ and $f(-1) = -1$, f is not one-to-one. Therefore f has no inverse.

25. Because $f \colon X \to X$ is a one-to-one correspondence, f has an inverse, which can be found by solving $y = f(x)$ for x in terms of y.

$$y = 3x - 6$$
$$y + 6 = 3x$$
$$\frac{y + 6}{3} = x$$

Therefore $f^{-1} \colon X \to X$ is defined by $f^{-1}(x) = \dfrac{x + 6}{3}$.

27. A sundae is determined by some subset of the five toppings. Hence the number of different sundaes equals the number of subsets of a set with five elements, which is $2^5 = 32$ by Theorem 1.3.

29. The number of different grievance committees equals the number of different ways to form a 6-element subset from a set of 15 people. This number is

$$C(15,6) = \frac{15!}{6!\,(15-6)!} = \frac{15 \cdot 14 \cdot 13 \cdot 12 \cdot 11 \cdot 10}{6 \cdot 5 \cdot 4 \cdot 3 \cdot 2 \cdot 1} = 5005.$$

31.   (i) Since $x - x = 0 \in \{-4, 0, 4\}$ for each $x \in S$, R is reflexive.

(ii) If $x - y \in \{-4, 0, 4\}$, then $y - x = -(x - y) \in \{-4, 0, 4\}$. Thus R is symmetric.

(iii) It is easy to see that 1 and 5 are the only elements related to 1 and 5, 2 and 6 are the only elements related to 2 and 6, 3 and 7 are the only elements related to 3 and 7, and 4 and 8 are the only elements related to 4 and 8. From this observation it is clear that R must be transitive.

From (iii) we see that R has four equivalence classes, which are

$$\{1,5\}, \qquad \{2,6\}, \qquad \{3,7\}, \qquad \text{and} \qquad \{4,8\}.$$

33.   (i) Since $x = x$ for every integer x, R is reflexive.

(ii) Suppose that $x\,R\,y$. Then either $x = y$ or $|x - y| = 1$ and the larger of x and y is even. It follows that either $y = x$ or $|y - x| = 1$ and the larger of y and x is even. Hence $y\,R\,x$, so that R is symmetric.

(iii) Suppose that $x\,R\,y$ and $y\,R\,z$. Then either $x = y$ or $|x - y| = 1$ and the larger of x and y is even, and either $y = z$ or $|y - z| = 1$ and the larger of y and z is even. When $x = y$ or $y = z$, it is easily checked that $x\,R\,z$. Suppose then that $x \neq y$

and $y \neq z$. If $x > y$, then x is even and $y = x - 1$ is odd. Thus we must have $z > y$ in order that the larger of y and z be even. Hence in this case we must have $x = z$. If $x < y$, a similar argument shows that $x = z$. Therefore, in each case, we have $x \ R \ z$, and so R is transitive.

Consider an even integer $2n$. Because $2n + 1$ is odd, the only integers x such that $x \ R \ 2n$ are $x = 2n$ and $x = 2n - 1$. Hence the equivalence class containing $2n$ is $\{2n - 1, 2n\}$. There is an equivalence class of this form for every integer n.

35. If a divides b, then there is an integer q such that $b = aq$. Let $y \in B$. Then $y \equiv x$ (mod b). Thus there is an integer m such that $y - x = mb$. It follows that $y - x = m(aq) = mqa$, so that a divides $y - x$. Hence $y \equiv x$ (mod a), and therefore $y \in A$. Since an arbitrary element of B is also an element of A, we have $B \subseteq A$.

37. By Theorem 2.4, there is a one-to-one correspondence between the equivalence relations on $S = \{a, b, c\}$ and the partitions of S. It is easily seen that there are exactly five partitions of S, namely,

(i) $\{\{a\}, \{b\}, \{c\}\}$

(ii) $\{\{a, b\}, \{c\}\}$

(iii) $\{\{a, c\}, \{b\}\}$

(iv) $\{\{a\}, \{b, c\}\}$

(v) $\{\{a, b, c\}\}$.

These partitions correspond to the following five equivalence relations on S:

(i) $\{(a, a), (b, b), (c, c)\}$

(ii) $\{(a, a), (a, b), (b, a), (b, b), (c, c)\}$

(iii) $\{(a, a), (a, c), (b, b), (c, a), (c, c)\}$

(iv) $\{(a, a), (b, b), (b, c), (c, b), (c, c)\}$

(v) $\{(a, a), (a, b), (a, c), (b, a), (b, b), (b, c), (c, a), (c, b), (c, c)\} = S \times S$.

39. Let R be an equivalence relation on set S that is also a function with domain S. Because R is reflexive, we must have $x \ R \ x$ for each $x \in S$. But, by definition of a function, there can be only one y such that $x \ R \ y$. Hence $R = \{(x, x): x \in S\}$.

41. Because $A \neq A$ is false for every subset A of $\{1, 2, 3, 4\}$, relation R is not reflexive. If $A \ R \ B$ is true, then $A \subseteq B$ and $A \neq B$. Hence $B \ R \ A$ is false. Therefore $A \ R \ B$ and $B \ R \ A$ is false for all subsets A and B of $\{1, 2, 3, 4\}$. Thus R is antisymmetric. Finally, suppose that both $A \ R \ B$ and $B \ R \ C$ are true. Then $A \subseteq B$, $A \neq B$, $B \subseteq C$, and $B \neq C$. Since the "is a subset of" relation is transitive, it follows that $A \subseteq C$ and $A \neq C$. Hence $A \ R \ C$, and so R is transitive.

43. Since $x = (1)^2 x$ for all $x \in S$, relation R is reflexive. Suppose that $x, y \in S$, and both $x \, R \, y$ and $y \, R \, x$ are true. Then $y = m^2 x$ for some integer m, and $x = n^2 y$ for some integer n. It follows that $y = m^2 x = m^2(n^2 y) = m^2 n^2 y$. Thus $m^2 = n^2 = 1$, and so $y = m^2 x = 1 \cdot x = x$. Therefore R is antisymmetric. Finally, let $x, y, z \in S$ be such that $x \, R \, y$ and $y \, R \, z$. Then $y = m^2 x$ for some integer m, and $z = n^2 y$ for some integer n. Hence $z = n^2 y = n^2(m^2 x) = (mn)^2 x$. Because mn is an integer if m and n are integers, it follows that $x \, R \, z$. Hence R is transitive.

45. In the creation of the advertisement for the Fourth of July sale, the precedence relations among the tasks establishes a partial order R on the set of tasks

$$S = \{\text{A, B, C, D, E, F, G, H, I, J, K}\}.$$

We must find a total order that contains R. From Figure 1.2(a), we obtain the Hasse diagram for R and S shown below.

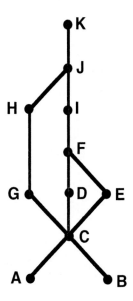

Applying the topological sorting algorithm to this situation, we obtain a total order T that contains R. We then perform task X before task Y if $(X, Y) \in T$. One possible sequencing of the tasks that results from this procedure is

$$\text{A, B, C, D, E, F, G, H, I, J, K.}$$

47. Let R be a relation on set S that is both an equivalence relation and a partial order. Because R is reflexive, $(x, x) \in R$ for each $x \in S$. Suppose that $(x, y) \in R$ for some $x, y \in S$. Since R is symmetric, it follows that $(y, x) \in R$. Because R is also antisymmetric, $(x, y) \in R$ and $(y, x) \in R$ imply that $x = y$. Therefore R is the equality relation on S, that is, $R = \{(x, x) : x \in S\}$.

49. Let R be a partial order on S, and let $x, y \in S$ be such that $x \vee y$ exists. Denote $x \vee y$ by z. Then $x \mathrel{R} z$, $y \mathrel{R} z$, and, if $w \in S$ and both $x \mathrel{R} w$ and $y \mathrel{R} w$, then $z \mathrel{R} w$. The preceding sentence is equivalent to $y \mathrel{R} z$, $x \mathrel{R} z$, and, if $w \in S$ and both $y \mathrel{R} w$ and $x \mathrel{R} w$, then $z \mathrel{R} w$. Hence $z = y \vee x$.

51. Let R be a partial order on S, and let x, y, and z be elements of S such that $x \vee y$, $y \vee z$, $(x \vee y) \vee z$, and $x \vee (y \vee z)$ all exist. Let $u = (x \vee y) \vee z$. Then $(x \vee y) \mathrel{R} u$ and $z \mathrel{R} u$. Now $(x \vee y) \mathrel{R} u$ implies $x \mathrel{R} u$ and $y \mathrel{R} u$. Since $y \mathrel{R} u$ and $z \mathrel{R} u$, we have $(y \vee z) \mathrel{R} u$. Hence both $x \mathrel{R} u$ and $(y \vee z) \mathrel{R} u$. Next suppose that $w \in S$ and $x \mathrel{R} w$ and $(y \vee z) \mathrel{R} w$. We will show that $u \mathrel{R} w$. Now $y \mathrel{R} (y \vee z)$, and because $(y \vee z) \mathrel{R} w$, the transitivity of R shows that $y \mathrel{R} w$. A similar argument shows that $z \mathrel{R} w$. Since $x \mathrel{R} w$ and $y \mathrel{R} w$, we have $(x \vee y) \mathrel{R} w$. But also $z \mathrel{R} w$, and so $u \mathrel{R} w$. It follows that $u = x \vee (y \vee z)$.

53. (i) Because $f(s) = f(s)$ for each $s \in Z$, R is reflexive.

(ii) If $x \mathrel{R} y$, then $f(x) = f(y)$. But then $f(y) = f(x)$; so $y \mathrel{R} x$, and R is symmetric.

(iii) If $x \mathrel{R} y$ and $y \mathrel{R} z$, then $f(x) = f(y)$ and $f(y) = f(z)$. Hence $f(x) = f(z)$, and so R is transitive.

55. If every equivalence class of f consists of a single element, then $f(n) = f(m)$ must imply that $n = m$, for m belongs to the equivalence class containing n. Hence f is one-to-one.

57. We have $(x, y) \in A \times (B \cup C)$ if and only if $x \in A$ and $y \in B \cup C$, which occurs if and only if $x \in A$, and $y \in B$ or $y \in C$ (or both), that is, if and only if $(x, y) \in (A \times B) \cup (A \times C)$. Therefore $A \times (B \cup C) = (A \times B) \cup (A \times C)$.

59. We have $(x, y) \in A \times (B - C)$ if and only if $x \in A$ and $y \in B - C$, if and only if $x \in A$, $y \in B$, and $y \notin C$, that is, if and only if $(x, y) \in (A \times B) - (A \times C)$. Therefore $A \times (B - C) = (A \times B) - (A \times C)$.

61. We have $x \in A \cap (C - B)$ if and only if $x \in A$ and $x \in C - B$, which happens if and only if $x \in A$, $x \in C$, and $x \notin B$. But these conditions occur if and only if $x \in A - B$ and $x \notin A - C$, that is, if and only if $x \in (A - B) - (A - C)$. Hence $A \cap (C - B) = (A - B) - (A - C)$.

63. For $n = 1$, the left side of the equation to be proved is 1 and the right side is

$$\frac{1(2 \cdot 1 - 1)(2 \cdot 1 + 1)}{3} = \frac{1 \cdot 1 \cdot 3}{3} = 1.$$

Thus the equation is true for $n = 1$. Assume that it is true for $n = k \geq 1$. Then

$$1^2 + 3^2 + \cdots + (2k-1)^2 + (2k+1)^2 = \left[1^2 + 3^2 + \cdots + (2k-1)^2\right] + (2k+1)^2$$

$$= \frac{k(2k-1)(2k+1)}{3} + (2k+1)^2$$

$$= \frac{k(2k-1)(2k+1) + 3(2k+1)^2}{3}$$

$$= \frac{[k(2k-1) + 3(2k+1)](2k+1)}{3}$$

$$= \frac{\left(2k^2 + 5k + 3\right)(2k+1)}{3}$$

$$= \frac{(k+1)(2k+3)(2k+1)}{3}$$

$$= \frac{(k+1)(2k+1)(2k+3)}{3}.$$

This proves the equation for $n = k+1$. Therefore the equation is correct for all positive integers n by the principle of mathematical induction.

65. If $n = 1$, the left side of the equation is

$$\frac{5}{1 \cdot 2 \cdot 3} = \frac{5}{6},$$

and the right side is

$$\frac{1(3 \cdot 1 + 7)}{2(1+1)(1+2)} = \frac{10}{2 \cdot 2 \cdot 3} = \frac{10}{12} = \frac{5}{6}.$$

Hence the equation is true for $n = 1$. Assume that the equation is true for some positive integer k. Then

$$\frac{5}{1 \cdot 2 \cdot 3} + \frac{6}{2 \cdot 3 \cdot 4} + \cdots + \frac{k+4}{k(k+1)(k+2)} + \frac{k+5}{(k+1)(k+2)(k+3)}$$

$$= \frac{k(3k+7)}{2(k+1)(k+2)} + \frac{k+5}{(k+1)(k+2)(k+3)}$$

$$= \frac{k(3k+7)(k+3)}{2(k+1)(k+2)(k+3)} + \frac{2(k+5)}{2(k+1)(k+2)(k+3)}$$

$$= \frac{3k^3 + 16k^2 + 23k + 10}{2(k+1)(k+2)(k+3)}$$

$$= \frac{(k+1)^2(3k+10)}{2(k+1)(k+2)(k+3)}$$

$$= \frac{(k+1)(3k+10)}{2(k+2)(k+3)}.$$

Thus the equation is true for $n = k+1$, and so it is true for all positive integers n by the principle of mathematical induction.

67. Postage of 8 cents can be made using one stamp worth 3 cents and one stamp worth 5 cents. Postage of 9 cents can be made using three stamps worth 3 cents each. Postage of 10 cents can be made using two stamps worth 5 cents each. Assume that for some integer $k \geq 10$, postage of 8 cents, 9 cents, ..., k cents can be made using only stamps worth 3 cents and 5 cents. Then $k - 2 \geq 8$; so postage of $k - 2$ cents can be made using only stamps worth 3 cents and 5 cents. By adding one stamp worth 3 cents to these stamps, we obtain postage worth $k + 1$ cents using only stamps worth 3 cents and 5 cents. This proves the result for $k + 1$. Therefore the result is true for all integers $n \geq 8$ by the strong principle of mathematical induction.

69. If $n = 3$, the n-sided polygon is a triangle. Since the sum of the interior angles of a triangle is $180° = 3(180°) - 360°$, the formula is correct in this case. Assume that the formula is correct for k-sided polygons, and consider $k + 1$ consecutive points $P_1, P_2, \ldots, P_{k+1}$ on the circumference of a circle. By the inductive hypothesis, the k-sided polygon determined by $P_1, P_2, \ldots, P_k$ has an angle sum of $k(180°) - 360°$. The interior angles of the $(k+1)$-sided polygon determined by $P_1, P_2, \ldots, P_{k+1}$ are the same as those of the k-sided polygon determined by $P_1, P_2, \ldots, P_k$ except for the angles at P_1, P_k, and P_{k+1}. (See the figure below.)

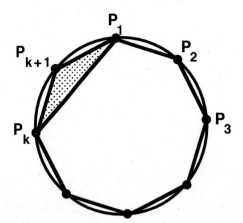

Since P_1, P_k, and P_{k+1} form a triangle, the amount by which the sum of the interior angles of the $(k+1)$-sided polygon exceeds the sum of the interior angles of the k-sided polygon is $180°$. Therefore the angle sum of the $(k+1)$-sided polygon is

$$[k(180°) - 360°] + 180° = (k+1)180° - 360°.$$

Thus the formula is correct for $k + 1$, and so it is correct for all integers $n \geq 3$ by the principle of mathematical induction.

71. For $n = 1$, $F_{mn} = F_m$ is clearly divisible by F_m. Assume that for some positive integer k, F_{mk} is divisible by F_m, say, $F_{mk} = r \cdot F_m$. Apply Exercise 70 with $n = mk + m - 1$ to obtain

$$F_{m(k+1)} = F_{mk-1}F_m + F_{mk}F_{m+1} = F_m(F_{mk-1} + rF_{m+1}).$$

Thus $F_{m(k+1)}$ is divisible by F_m. Therefore F_{mn} is divisible by F_m for all positive integers n by the principle of mathematical induction.

Chapter 3

Coding Theory

3.1 CONGRUENCE

1. Since $67 = 7 \cdot 9 + 4$, the quotient is 7 and the remainder is 4.

3. Since $25 = 0 \cdot 42 + 25$, the quotient is 0 and the remainder is 25.

5. Since $-54 = -9 \cdot 6 + 0$, the quotient is -9 and the remainder is 0.

7. Since $-89 = 9(-10) + 1$, the quotient is -9 and the remainder is 1.

9. Since $p - q = 63$ is divisible by m, $p \equiv q \pmod{m}$.

11. Since $p - q = 61$ is not divisible by m, $p \not\equiv q \pmod{m}$.

13. Since $p - q = 85$ is not divisible by m, $p \not\equiv q \pmod{m}$.

15. Since $p - q = 54$ is divisible by m, $p \equiv q \pmod{m}$.

17. In $\mathcal{Z}_{12}$, $[8] + [6] = [14] = [2]$ because 2 is the remainder when 14 is divided by 12.

19. In $\mathcal{Z}_{11}$, $[5] + [10] = [15] = [4]$ since 4 is the remainder when 15 is divided by 11.

21. In $\mathcal{Z}_8$, $[23] + [15] = [38] = [6]$ because 6 is the remainder when 38 is divided by 8.

23. In $\mathcal{Z}_6$, $[16] + [9] = [25] = [1]$.

25. In $\mathcal{Z}_6$, $[8][7] = [56] = [2]$.

27. In $\mathcal{Z}_9$, $[4][11] = [44] = [8]$.

29. In $\mathcal{Z}_8$, $[5][12] = [60] = [4]$.

31. In $\mathcal{Z}_{10}$, $[9][6] = [54] = [4]$.

33. In $\mathcal{Z}_7$, $[9]^7 = [2]^7 = [2^7] = [128] = [2]$.

35. In $\mathcal{Z}_{12}$, $[11]^9 = [-1]^9 = [(-1)^9] = [-1] = [11]$.

37. Since the machine uses 4 feet of paper per hour, a new roll of paper lasts 50 hours. The eighteenth hour of the day is 6 P.M., and $[18] + [50] = [68] = [20]$ in $\mathcal{Z}_{24}$. Thus the paper supply runs out at the twentieth hour of the day (at 8 P.M.).

39. Since

$$10(3) + 9(5) + 8(4) + 7(0) + 6(9) + 5(0) + 4(5) + 3(1) + 2(8) = 200$$

and

$$200 + 9 = 209 = 19 \cdot 11 \equiv 0 \ (\text{mod } 11),$$

the check digit is 9.

41. We have

$$3(0) + 1(7) + 3(0) + 1(3) + 3(3) + 1(0) + 3(2) + 1(0) + 3(1) + 1(1) + 3(8) = 53.$$

The check digit c is chosen so that $53 + c \equiv 0 \ (\text{mod } 10)$. Thus the check digit is 7.

43. **(a)** Using the formula in Example 3.7 with $x = 2020$, we obtain

$$
\begin{aligned}
y &\equiv x + f(x-1) - g(x-1) + h(x-1) \\
&\equiv 2020 + f(2019) - g(2019) + h(2019) \\
&\equiv 2020 + 5 - 20 + 504 \\
&\equiv 2509 \\
&\equiv 3 \ (\text{mod } 7).
\end{aligned}
$$

Therefore January 1, 2020 falls on a Wednesday. Now February 12, 2020 occurs $31 + (12 - 1) = 42$ days after January 1, 2020. Because

$$3 + 42 \equiv 45 \equiv 3 \ (\text{mod } 7),$$

February 12, 2020 also falls on a Wednesday.

(b) The year 2020 is a leap year. Hence August 8, 2020 occurs

$$31 + 29 + 31 + 30 + 31 + 30 + 31 + (8 - 1) = 220$$

days after January 1, 2020. Now

$$3 + 220 \equiv 223 \equiv 6 \ (\text{mod } 7),$$

and so August 8, 2020 falls on a Saturday.

45. The sets A and B are not equal because $10 \in A$ but $10 \notin B$.

47. For this equivalence relation R, both $3\ R\ 11$ and $6\ R\ 10$ are true, but both $9\ R\ 21$ and $18\ R\ 110$ are false.

49.   (i) We must first show that this relation is well-defined on $\mathcal{Z}_n$, that is, if $[x] = [x']$ and $[y] = [y']$, then $[x]\ R\ [y]$ is true if and only if $[x']\ R\ [y']$ is true. Let $[x] = [x']$ and $[y] = [y']$. Then $x' = x + un$ and $y' = y + vn$ for some integers u and v. Now $[x']\ R\ [y']$ if and only if n divides $x' - y'$, if and only if $x' - y' = kn$ for some integer k, if and only if $(x + un) - (y + vn) = kn$ for some k, if and only if $x - y = (v - u + k)n$ for some k, if and only if n divides $x - y$, if and only if $[x]\ R\ [y]$. So R is well-defined.

   (ii) If $[x] \in \mathcal{Z}_n$, then $[x]\ R\ [x]$ because $x \equiv x \pmod{m}$. Thus R is reflexive.

   (iii) Let $[x]\ R\ [y]$. Then $x \equiv y \pmod{m}$, and so $y \equiv x \pmod{m}$. Hence $[y]\ R\ [x]$, and thus R is symmetric.

   (iv) Let $[x]\ R\ [y]$ and $[y]\ R\ [z]$. Then $x \equiv y \pmod{m}$ and $y \equiv z \pmod{m}$. Therefore $x \equiv z \pmod{m}$, and so $[x]\ R\ [z]$. Thus R is transitive.

   (v) If m does not divide n, then this relation is not well-defined, for $[0] = [n]$, but $0 \not\equiv n \pmod{m}$.

51.   **(a)** We have $x \equiv y \pmod{m}$ if and only if $x - y = km$ for some integer k, if and only if $x = km + y$ for some integer k.

   (b) Suppose that x and y have the same remainder r when divided by m. Then $x = am + r$ and $y = bm + r$ for some integers a and b. Thus

$$x - y = (am + r) - (bm + r) = (a - b)m;$$

so $x \equiv y \pmod{m}$. Conversely suppose that $x \equiv y \pmod{m}$. Then by (a), $x = km + y$ for some integer k. Let q and r denote the quotient and remainder, respectively, when y is divided by m. Thus $y = qm + r$. But then

$$x = km + y = km + (qm + r) = (k + q)m + r,$$

so that $k + q$ is the quotient and r is the remainder when x is divided by m. Hence x and y have the same remainder r when divided by m.

3.2   THE EUCLIDEAN ALGORITHM

1. By trying each, we find the positive divisors of 45 to be 1, 3, 5, 9, 15, and 45. Thus all divisors, in order, are $-45, -15, -9, -5, -3, -1, 1, 3, 5, 9, 15, 45$.

3. By trying each, we find the positive divisors of 40 to be 1, 2, 4, 5, 8, 10, 20, 40, and the positive divisors of 30 to be 1, 2, 3, 5, 6, 10, 15, 30. The common positive divisors are 1, 2, 5, and 10. Thus all common divisors, in order, are $-10, -5, -2, -1, 1, 2, 5, 10$.

5. We set $r_{-1} = 715$, $r_0 = 312$, and $i = 0$. Since $r_0 \neq 0$, we replace i with 1. Dividing 715 by 312 gives $715 = 2 \cdot 312 + 91$; so we set $r_1 = 91$. Continuing in this way gives the following table.

i	r_i
-1	715
0	312
1	91
2	39
3	13
4	0

Since $r_4 = 0$, we print $r_3 = 13$.

7. We set $r_{-1} = 247$, $r_0 = 117$, and $i = 0$. Since $r_0 \neq 0$, we replace i with 1. Dividing 247 by 117 gives $247 = 2 \cdot 117 + 13$; so we set $r_1 = 13$. Continuing in this way gives the following table.

i	r_i
-1	247
0	117
1	13
2	0

Since $r_2 = 0$, we print $r_1 = 13$.

9. We set $r_{-1} = 76$, $r_0 = 123$, and $i = 0$. Since $r_0 \neq 0$, we replace i with 1. Dividing 76 by 123 gives $76 = 0 \cdot 123 + 123$; so we set $r_1 = 76$. Continuing in this way gives the following table.

i	r_i
-1	76
0	123
1	76
2	47
3	29
4	18
5	11
6	7
7	4
8	3
9	1
10	0

Since $r_{10} = 0$, we print $r_9 = 11$.

11. By Theorem 3.4, the number of divisions needed is at most

$$2 \log_2(n+1) \leq 2 \log_2(999999 + 1) = 2 \log_2 1000000 = 2 \, \frac{\log 1000000}{\log 2} \approx 39.9,$$

where we have used the formula $\log_2 x = \log x / \log 2$ of Section 2.5. Thus at most 39 divisions are needed.

13. We set $r_{-1} = 1479$, $x_{-1} = 1$, $y_{-1} = 0$, $r_0 = 272$, $x_0 = 0$, $y_0 = 1$, and $i = 0$. Since $r_0 \neq 0$, we replace i with 1. Dividing r_{-1} by r_0 gives $1479 = 5 \cdot 272 + 119$; so $q_1 = 5$ and $r_1 = 119$. Then we set $x_1 = 1 - 5 \cdot 0 = 1$ and $y_1 = 0 - 5 \cdot 1 = -5$. Continuing in this way leads to the following table.

i	q_i	r_i	x_i	y_i
-1		1479	1	0
0		272	0	1
1	5	119	1	-5
2	2	34	-2	11
3	3	17	7	-38
4	2	0	-16	87

Since $r_4 = 0$, we print $r_3 = 17$, $x_3 = 7$, and $y_3 = -38$.

15. We set $r_{-1} = 4050$, $x_{-1} = 1$, $y_{-1} = 0$, $r_0 = 1728$, $x_0 = 0$, $y_0 = 1$, and $i = 0$. Since $r_0 \neq 0$, we replace i with 1. Dividing r_{-1} by r_0 gives $4050 = 2 \cdot 1728 + 594$; so $q_1 = 2$ and $r_1 = 594$. Then we set $x_1 = 1 - 2 \cdot 0 = 1$ and $y_1 = 0 - 2 \cdot 1 = -2$. Continuing in this way leads to the following table.

i	q_i	r_i	x_i	y_i
-1		4050	1	0
0		1728	0	1
1	2	594	1	-2
2	2	540	-2	5
3	1	54	3	-7
4	10	0	-32	75

Since $r_4 = 0$, we print $r_3 = 54$, $x_3 = 3$, and $y_3 = -7$.

17. We set $r_{-1} = 546$, $x_{-1} = 1$, $y_{-1} = 0$, $r_0 = 2022$, $x_0 = 0$, $y_0 = 1$, and $i = 0$. Since $r_0 \neq 0$, we replace i with 1. Dividing r_{-1} by r_0 gives $546 = 0 \cdot 2022 + 546$; so $q_1 = 0$ and $r_1 = 546$. Then we set $x_1 = 1 - 0 \cdot 0 = 1$ and $y_1 = 0 - 0 \cdot 1 = 0$. Continuing in this way leads to the following table.

i	q_i	r_i	x_i	y_i
-1		546	1	0
0		2022	0	1
1	0	546	1	0
2	3	384	-3	1
3	1	162	4	-1
4	2	60	-11	3
5	2	42	26	-7
6	1	18	-37	10
7	2	6	100	-27
8	3	0	-337	91

Since $r_8 = 0$, we print $r_7 = 6$, $x_7 = 100$, and $y_7 = -27$.

19. Using the Euclidean algorithm, we find that $\gcd(414, 594) = 18$. Since 18 does not divide 492, the equation in (a) is not solvable, since for any integers x and y, the left side is divisible by 18. On the other hand, $558 = 18 \cdot 31$. Thus we could use the extended Euclidean algorithm to find integers x and y such that $414x + 594y = 18$. But then $414(31x) + 594(31y) = 18 \cdot 31 = 558$. Thus the equation in (b) is solvable.

21. Using the Euclidean algorithm, we find that $\gcd(396, 312) = 12$. Since 12 does not divide 222, the equation in (a) is not solvable, since for any integers x and y, the left side is divisible by 12. On the other hand, $228 = 12 \cdot 19$. Thus we could use the extended Euclidean algorithm to find integers x and y such that $396x + 312y = 12$. But then $396(19x) + 312(19y) = 12 \cdot 19 = 228$. Thus the equation in (b) is solvable.

23. Applying the Euclidean algorithm to 3157 and 656 reveals that $\gcd(3157, 656) = 41$. Now $2173 = 41 \cdot 53$; so solutions exist. Using the extended Euclidean algorithm gives the following table.

i	q_i	r_i	x_i	y_i
-1		3157	1	0
0		656	0	1
1	4	533	1	-4
2	1	123	-1	5
3	4	41	5	-24
4	3	0	-16	77

Since $r_4 = 0$, we have $r_3 = 41$ (which we already knew), $x_3 = 5$, and $y_3 = -24$. Thus $3157 \cdot 5 + 656(-24) = 41$. (Check this!) Now we need only to multiply this equation by 53. We have $3157 \cdot (53 \cdot 5) + 656(53(-24)) = 53 \cdot 41 = 2173$. Thus a solution is $x = 53 \cdot 5 = 265$ and $y = 53(-24) = -1272$. (There are other solutions.)

25. Since the Euclidean algorithm and extended Euclidean algorithm specify nonnegative integers, we will start with $m = 455$ and $n = 169$ and worry about the minus sign later.

45

By the Euclidean algorithm, $\gcd(455, 169) = 13$. Now $1157 = 13 \cdot 89$; so solutions exist. Using the extended Euclidean algorithm with the same m and n gives the following table.

i	q_i	r_i	x_i	y_i
-1		455	1	0
0		169	0	1
1	2	117	1	-2
2	1	52	-1	3
3	2	13	3	-8
4	4	0	-13	35

Since $r_4 = 0$, we have $r_3 = 13$ (which we already knew), $x_3 = 3$, and $y_3 = -8$. Thus $455 \cdot 3 + 169(-8) = 13$. (Check this!) We multiply this equation by 89 to find $455 \cdot (89 \cdot 3) + 169(89(-8)) = 455 \cdot 267 + 169(-712) = 89 \cdot 13 = 1157$. To get the minus sign of the original equation, we write this as $455 \cdot 267 - 169 \cdot 712 = 1157$. Thus a solution is $x = 267$ and $y = 712$. (There are other solutions.)

3.3 THE RSA METHOD

1. We have

$$2^5 = 32 = 0 \cdot 35 + 32,$$
$$5^5 = 3125 = 89 \cdot 35 + 10,$$
$$11^5 = 161051 = 4601 \cdot 3516,$$
$$8^5 = 32768 = 936 \cdot 35 + 8.$$

Thus the ciphertext is 32, 10, 16, 8.

3. We have

$$40^3 = 64000 = 1163 \cdot 55 + 35,$$
$$31^3 = 29791 = 541 \cdot 55 + 36,$$
$$9^3 = 729 = 13 \cdot 55 + 14.$$

Thus the ciphertext is 35, 36, 14.

5. We start by setting $r_2 = 1$, $p = 19$, and $e = 41$. Since $e \neq 0$, we write $41 = 2 \cdot 20 + 1$; so $Q = 20$ and $R = 1$. Also $p^2 = 19^2 = 361 = 3 \cdot 91 + 88$; so $r_1 = 88$. Since $R = 1$, we write $r_2 p = 1 \cdot 19 = 0 \cdot 91 + 19$, and replace r_2 with 19. Then we set $p = 88$ and $e = 20$. Thus far we have the following table.

Q	R	r_1	r_2	p	e
			1	19	41
20	1	88	19	88	20

Continuing with the algorithm leads to the following.

Q	R	r_1	r_2	p	e
			1	19	41
20	1	88	19	88	20
10	0	9		9	10
5	0	81		81	5
2	1	9	83	9	2
1	0	81		81	1
0	1	9	80	9	0

Since now $e = 0$, we print $r_2 = 80$.

7. We start by setting $r_2 = 1$, $p = 11$, and $e = 73$. Since $e \neq 0$, we write $73 = 2 \cdot 36 + 1$; so $Q = 36$ and $R = 1$. Also $p^2 = 11^2 = 121 = 0 \cdot 187 + 121$; so $r_1 = 121$. Since $R = 1$, we write $r_2 p = 1 \cdot 11 = 0 \cdot 187 + 11$, and replace r_2 with 11. Then we set $p = 121$ and $e = 36$. Thus far we have the following table.

Q	R	r_1	r_2	p	e
			1	11	73
36	1	121	11	121	36

Continuing with the algorithm leads to the following.

Q	R	r_1	r_2	p	e
			1	11	73
36	1	121	11	121	36
18	0	55		55	18
9	0	33		33	9
4	1	154	176	154	4
2	0	154		154	2
1	0	154		154	1
0	1	154	176	154	0

Since now $e = 0$, we print $r_2 = 176$.

9. We start by setting $r_2 = 1$, $p = 90$, and $e = 101$. Since $e \neq 0$, we write $101 = 2 \cdot 50 + 1$; so $Q = 50$ and $R = 1$. Also $p^2 = 90^2 \equiv 966 \pmod{1189}$ by the text following Example

47

3.11; so $r_1 = 966$. Since $R = 1$, we write $r_2p = 1 \cdot 90 = 0 \cdot 1189 + 90$, and replace r_2 with 90. Then we set $p = 966$ and $e = 50$. Thus far we have the following table.

Q	R	r_1	r_2	p	e
			1	90	101
50	1	966	90	966	50

Continuing with the algorithm leads to the following.

Q	R	r_1	r_2	p	e
			1	90	101
50	1	966	90	966	50
25	0	980		980	25
12	1	877	214	877	12
6	0	1035		1035	6
3	0	1125		1125	3
1	1	529	572	529	1
0	1	426	582	426	0

Since now $e = 0$, we print $r_2 = 582$.

11. Since $85 = 5 \cdot 17$, we take $p = 5$ and $q = 17$. Then $b = (5-1)(17-1) = 64$.

13. Since $323 = 17 \cdot 19$, we take $p = 17$ and $q = 19$. Then $b = (17-1)(19-1) = 288$.

15. Since $35 = 5 \cdot 7$, $b = (5-1)(7-1) = 24$. We must find the smallest positive solution to $5x \equiv 1 \pmod{24}$. Taking $m = 24$ and $n = 5$ in the extended Euclidean algorithm leads to the following table.

i	q_i	r_i	x_i	y_i
-1		24	1	0
0		5	0	1
1	4	4	1	-4
2	1	1	-1	5
3	4	0	5	-24

Thus $24(-1) + 5 \cdot 5 = 1$, and $5 \cdot 5 = 1 + 24 \equiv 1 \pmod{24}$. Since $0 \leq 5 < 24$, we take $D = 5$.

17. Since $55 = 5 \cdot 11$, $b = (5-1)(11-1) = 40$. We must find the smallest positive solution to $3x \equiv 1 \pmod{40}$. Taking $m = 40$ and $n = 3$ in the extended Euclidean algorithm leads to the following table.

i	q_i	r_i	x_i	y_i
-1		40	1	0
0		3	0	1
1	13	1	1	-13
2	3	0	-3	40

Thus $40 \cdot 1 + 3(-13) = 1$, and $3(-13) = 1 - 40 \equiv 1 \pmod{40}$. Since

$$-13 = -1 \cdot 40 + 27 \qquad \text{and} \qquad 0 \le 27 < 40,$$

we take $D = 27$.

19. Since $91 = 7 \cdot 13$, $b = (7-1)(13-1) = 72$. We must find the smallest positive solution to $41x \equiv 1 \pmod{72}$. Taking $m = 72$ and $n = 41$ in the extended Euclidean algorithm leads to the following table.

i	q_i	r_i	x_i	y_i
-1		72	1	0
0		41	0	1
1	1	31	1	-1
2	1	10	-1	2
3	3	1	4	-7
4	10	0	-41	72

Thus $72 \cdot 4 + 41(-7) = 1$, and $41(-7) = 1 - 4 \cdot 72 \equiv 1 \pmod{72}$. Since

$$-7 = -1 \cdot 72 + 65 \qquad \text{and} \qquad 0 \le 65 < 72,$$

we take $D = 65$.

21. Since $187 = 11 \cdot 17$, $b = (11-1)(17-1) = 160$. We must find the smallest positive solution to $73x \equiv 1 \pmod{160}$. Taking $m = 160$ and $n = 73$ in the extended Euclidean algorithm leads to the following table.

i	q_i	r_i	x_i	y_i
-1		160	1	0
0		73	0	1
1	2	14	1	-2
2	5	3	-5	11
3	4	2	21	-46
4	1	1	-26	57
5	2	0	73	-160

Thus $160(-26) + 73 \cdot 57 = 1$, and $73 \cdot 57 = 1 + 1160 \cdot 26 \equiv 1 \pmod{160}$. Since $0 \le 57 < 160$, we take $D = 57$.

23. Since $55 = 5 \cdot 11$, $b = (5-1)(11-1) = 40$. We must find the smallest positive solution to $7x \equiv 1 \pmod{40}$. Taking $m = 40$ and $n = 7$ in the extended Euclidean algorithm leads to the following table.

i	q_i	r_i	x_i	y_i
-1		40	1	0
0		7	0	1
1	5	5	1	-5
2	1	2	-1	6
3	2	1	3	-17
4	2	0	-7	40

Thus $40 \cdot 1 + 7(-17) = 1$, and $7(-17) = 1 - 40 \equiv 1 \pmod{40}$. Since

$$-17 = (-1)40 + 23 \qquad \text{and} \qquad 0 \le 23 < 40,$$

we take $D = 23$.

Now we need to compute the remainder when 2^{23} is divided by 55. We use the modular exponentiation algorithm with $P = 2$, $E = 23$, and $n = 55$, in the notation of that algorithm, getting the following table.

Q	R	r_1	r_2	p	e
			1	2	23
11	1	4	2	4	11
5	1	16	8	16	5
2	1	36	18	36	2
1	0	31		31	1
0	1	26	8	26	0

We see that the remainder when 2^{23} is divided by 55 is 8. Thus, in the notation of this problem, $P = 8$.

3.4 ERROR-DETECTING AND ERROR-CORRECTING CODES

1. The codeword 01001010 contains an odd number of 1s (in the second, fifth, and seventh digits). Hence a parity check digit of 1 must be appended to make the total number of 1s even.

3. Because the given codeword contains exactly four 1s, a parity check digit of 0 must be appended to make the total number of 1s even.

5. Because the given codeword contains exactly four 1s, a parity check digit of 0 must be appended to make the total number of 1s even.

7. Because the given codeword contains exactly three 1s, a parity check digit of 1 must be appended to make the total number of 1s even.

9. Using formula (3.1), we obtain a probability of

$$C(n,p)p^k(1-p)^{n-k} = C(5,1)(.01)^1(.99)^4 = 5(.01)^1(.99)^4 \approx .0480.$$

11. Using formula (3.1), we obtain a probability of

$$C(n,p)p^k(1-p)^{n-k} = C(7,2)(.01)^2(.99)^5 = 21(.01)^2(.99)^5 \approx .0020.$$

13. Using formula (3.1), we obtain a probability of

$$C(n,p)p^k(1-p)^{n-k} = C(8,0)(.01)^0(.99)^8 = 1(.99)^8 \approx .9227.$$

15. Using formula (3.1), we obtain a probability of

$$C(n,p)p^k(1-p)^{n-k} = C(10,1)(.01)^1(.99)^9 = 10(.01)^1(.99)^9 \approx .0914.$$

17. Codewords c_1 and c_2 differ in their first and second digits only. Hence the Hamming distance between them is 2.

19. Codewords c_1 and c_2 differ in their first, third, fourth, and fifth digits only. Hence the Hamming distance between them is 4.

21. Codewords c_1 and c_2 differ in their first and second digits only. Hence the Hamming distance between them is 2.

23. Codewords c_1 and c_2 differ in their first, third, sixth, and seventh digits only. Hence the Hamming distance between them is 4.

25. When adding over $\mathcal{Z}_2$, we add corresponding digits using the rules

$$0+0=0, \qquad 0+1=1, \qquad 1+0=1, \qquad 1+1=0.$$

Thus the sum of c_1 and c_2 is 1100.

27. As in Exercise 25, the sum of c_1 and c_2 is 00101.

29. As in Exercise 25, the sum of c_1 and c_2 is 100101.

31. As in Exercise 25, the sum of c_1 and c_2 is 11010110.

33. According to Theorem 3.4, if m is the minimal Hamming distance between codewords, then $m-1$ or fewer errors can be detected, and $(m-1)/2$ or fewer errors can be corrected.

    (a) If $m=8$, then at most $8-1=7$ errors can be detected.
    (b) If $m=8$, then

$$\frac{m-1}{2} = \frac{7}{2} = 3.5.$$

Because the number of errors that can be corrected must be an integer, we see that at most 3 errors can be corrected.

35. **(a)** At most $15 - 1 = 14$ errors can be detected.

(b) At most
$$\frac{15 - 1}{2} = \frac{14}{2} = 7$$

errors can be corrected.

37. Let c_1 and c_2 be codewords of the same length. Corresponding digits of c_1 and c_2 differ if and only if the sum of these digits over $\mathcal{Z}_2$ is 1. Hence the Hamming distance between c_1 and c_2, which is the number of digits in which c_1 and c_2 differ, equals the number of 1s appearing in the sum of c_1 and c_2 over $\mathcal{Z}_2$.

39. The codewords corresponding to the 16 possible message words in $\mathcal{W}_4$ are:

00000000,	00010101,	00100110,	00110011,
01001001,	01011100,	01101111,	01111010,
10001010,	10011111,	10101100,	10111001,
11000011,	11010110,	11100101,	11110000.

It can be shown that the minimal Hamming distance between two of these codewords is 3. Thus, by Theorem 3.4, this code can correct all errors in a single digit.

41. According to Exercise 40, for every positive integer s, there exists an $(s^2, s^2 + 2s)$-block code that corrects all errors in a single digit. The efficiency of this code is
$$\frac{s^2}{s^2 + 2s} > \frac{s^2}{s^2 + 2s + 1} = \frac{s^2}{(s+1)^2} = \left(\frac{s}{s+1}\right)^2.$$

Given any ϵ such that $0 < \epsilon < 1$, choose s sufficiently large that
$$s > \frac{\sqrt{1 - \epsilon}}{1 - \sqrt{1 - \epsilon}}.$$

Then
$$\frac{1 - \sqrt{1 - \epsilon}}{\sqrt{1 - \epsilon}} > \frac{1}{s}$$

$$\frac{1}{\sqrt{1 - \epsilon}} - 1 > \frac{1}{s}$$

$$\frac{1}{\sqrt{1 - \epsilon}} > 1 + \frac{1}{s} = \frac{s + 1}{s}$$

$$s > \sqrt{1 - \epsilon}(s + 1)$$

$$\frac{s}{s + 1} > \sqrt{1 - \epsilon}$$

$$\left(\frac{s}{s + 1}\right)^2 > 1 - \epsilon.$$

Thus the code in Exercise 40 has efficiency greater than $1 - \epsilon$ and corrects all errors in a single digit.

3.5 MATRIX CODES

1. The number of words in $\mathcal{W}_5$ is $2^5 = 32$.

3. The number of words in $\mathcal{W}_8$ is $2^8 = 256$.

5. Let A denote the generator matrix. The codeword corresponding to a word w is wA. For $w = [1 \ 0 \ 0 \ 1]$, we have $wA = [1 \ 0 \ 0 \ 1 \ 0 \ 0 \ 1 \ 1]$. Thus the codeword associated with 1001 is 10010011. Notice that this is the sum over $\mathcal{Z}_2$ of the first and fourth rows of A.

7. The codeword associated with 1101 is 11010110, the sum over $\mathcal{Z}_2$ of the first, second, and fourth rows of A.

9. The generator matrix of a (3, 9)-block code is a 3×9 matrix. The corresponding check matrix is a $9 \times (9 - 3)$, that is, a 9×6, matrix.

11. The generator matrix of a (5, 10)-block code is a 5×10 matrix. The corresponding check matrix is a $10 \times (10 - 5)$, that is, a 10×5, matrix.

13. If the check matrix is a 9×3 matrix, then the generator matrix is a 6×9 matrix. Hence the efficiency of the code is $6/9 = 2/3$.

15. The given matrix A is a generator matrix for a (2, 5)-block code. The set of codewords for this code is

$$\{wA \colon w \in \mathcal{W}_2\} = \{00000, 10101, 01110, 11011\}.$$

17. The given matrix A is a generator matrix for a (3, 6)-block code. The set of codewords for this code is

$$\{wA \colon w \in \mathcal{W}_3\} = \{000000, 100001, 010011, 001111, 110010, 101110, 011100, 111101\}.$$

19. The given matrix A is a generator matrix for a (3, 7)-block code. The set of codewords for this code is

$$\{0000000, 1001001, 0100110, 0010101, 1101111, 1011100, 0110011, 1111010\}.$$

21. The given generator matrix has the form $[I_2|J]$, where

$$J = \begin{bmatrix} 0 & 1 & 1 \\ 1 & 1 & 1 \end{bmatrix}.$$

Therefore the check matrix is

$$\left[\frac{J}{I_3} \right] = \begin{bmatrix} 0 & 1 & 1 \\ 1 & 1 & 1 \\ 1 & 0 & 0 \\ 0 & 1 & 0 \\ 0 & 0 & 1 \end{bmatrix}.$$

23. The given generator matrix has the form $[I_3|J]$, where

$$J = \begin{bmatrix} 1 & 0 & 1 \\ 1 & 1 & 0 \\ 0 & 1 & 1 \end{bmatrix}.$$

Therefore the check matrix is

$$\left[\frac{J}{I_3} \right] = \begin{bmatrix} 1 & 0 & 1 \\ 1 & 1 & 0 \\ 0 & 1 & 1 \\ 1 & 0 & 0 \\ 0 & 1 & 0 \\ 0 & 0 & 1 \end{bmatrix}.$$

25. The given generator matrix has the form $[I_3|J]$, where

$$J = \begin{bmatrix} 0 & 1 & 1 & 1 \\ 1 & 0 & 1 & 1 \\ 1 & 1 & 0 & 1 \end{bmatrix}.$$

Thus the associated check matrix is

$$\left[\frac{J}{I_4} \right] = \begin{bmatrix} 0 & 1 & 1 & 1 \\ 1 & 0 & 1 & 1 \\ 1 & 1 & 0 & 1 \\ 1 & 0 & 0 & 0 \\ 0 & 1 & 0 & 0 \\ 0 & 0 & 1 & 0 \\ 0 & 0 & 0 & 1 \end{bmatrix}.$$

27. The given generator matrix has the form $[I_3|J]$, where

$$J = \begin{bmatrix} 1 & 0 & 1 & 0 & 1 \\ 0 & 1 & 0 & 1 & 0 \\ 0 & 1 & 1 & 1 & 0 \end{bmatrix}.$$

Thus the associated check matrix is

$$\left[\begin{array}{c} J \\ \hline I_5 \end{array}\right] = \begin{bmatrix} 1 & 0 & 1 & 0 & 1 \\ 0 & 1 & 0 & 1 & 0 \\ 0 & 1 & 1 & 1 & 0 \\ 1 & 0 & 0 & 0 & 0 \\ 0 & 1 & 0 & 0 & 0 \\ 0 & 0 & 1 & 0 & 0 \\ 0 & 0 & 0 & 1 & 0 \\ 0 & 0 & 0 & 0 & 1 \end{bmatrix}.$$

29. The given check matrix has the form $\left[\begin{array}{c} J \\ \hline I_3 \end{array}\right]$, where

$$J = \begin{bmatrix} 1 & 0 & 1 \\ 1 & 1 & 1 \\ 0 & 1 & 1 \\ 1 & 1 & 0 \end{bmatrix}.$$

Therefore the generator matrix is

$$[I_4 | J] = \begin{bmatrix} 1 & 0 & 0 & 0 & 1 & 0 & 1 \\ 0 & 1 & 0 & 0 & 1 & 1 & 1 \\ 0 & 0 & 1 & 0 & 0 & 1 & 1 \\ 0 & 0 & 0 & 1 & 1 & 1 & 0 \end{bmatrix}.$$

31. The given generator matrix A has the form $[I_3 | J]$, where

$$J = \begin{bmatrix} 1 & 0 & 1 & 1 \\ 0 & 1 & 1 & 0 \\ 0 & 0 & 1 & 1 \end{bmatrix}.$$

Thus the associated check matrix is

$$A^* = \left[\begin{array}{c} J \\ \hline I_4 \end{array}\right] = \begin{bmatrix} 1 & 0 & 1 & 1 \\ 0 & 1 & 1 & 0 \\ 0 & 0 & 1 & 1 \\ 1 & 0 & 0 & 0 \\ 0 & 1 & 0 & 0 \\ 0 & 0 & 1 & 0 \\ 0 & 0 & 0 & 1 \end{bmatrix}.$$

By Theorem 3.5(b), the given word $c = 0110110$ is a codeword for this code if and only if cA^* equals $[0\ 0\ 0\ 0]$. In this case, however, $cA^* = [0\ 0\ 1\ 1]$, and so c is *not* a codeword.

33. Let c denote the given codeword and A^* denote the check matrix in the solution to Exercise 31. Because $cA^* = [0\ \ 0\ \ 0\ \ 0]$, c is a codeword.

35. Let c denote the given codeword and A^* denote the check matrix in the solution to Exercise 31. Because $cA^* = [0\ \ 0\ \ 0\ \ 0]$, c is a codeword.

37. Let c denote the given codeword and A^* denote the check matrix in the solution to Exercise 31. Because $cA^* = [0\ \ 1\ \ 1\ \ 1]$, c is *not* a codeword.

39. In a (k, n)-code, the message words come from $\mathcal{W}_k$ and the codewords come from $\mathcal{W}_n$. Moreover, each message word corresponds uniquely to a codeword. Since $|\mathcal{W}_k| = 2^k$ and $|\mathcal{W}_n| = 2^n$, the proportion of words in $\mathcal{W}_n$ that are codewords is

$$\frac{2^k}{2^n} = 2^{n-k}.$$

41. From the solution to Exercise 39 in Section 3.4, we see that the codewords associated with the message words 1000, 0100, 0010, and 0001 are

$$10001010, \qquad 01001001, \qquad 00100110, \qquad \text{and} \qquad 00010101,$$

respectively. Thus these are the rows of the generator matrix, that is, the generator matrix is

$$\begin{bmatrix} 1 & 0 & 0 & 0 & 1 & 0 & 1 & 0 \\ 0 & 1 & 0 & 0 & 1 & 0 & 0 & 1 \\ 0 & 0 & 1 & 0 & 0 & 1 & 1 & 0 \\ 0 & 0 & 0 & 1 & 0 & 1 & 0 & 1 \end{bmatrix}.$$

43. Let e_r denote the $1 \times k$ matrix in which all the entries equal 0 except for entry r, which is 1. Then $e_r A$ and $e_r A'$ equal the rth rows of A and A', respectively. Now $e_r A'$ is a codeword in $C = C'$, and its first k digits are the same as e_r. But only one codeword in C has its first k digits the same as e_r, and this codeword is $e_r A$. Thus $e_r A' = e_r A$, that is, the rth rows of A and A' are equal. Because this is true for $r = 1, 2, \ldots, k$, it follows that $A = A'$.

3.6 MATRIX CODES THAT CORRECT ALL SINGLE-DIGIT ERRORS

1. The syndrome of the given word w is $wA^* = [0\ \ 1\ \ 1\ \ 0]$, which is the second row of A^*. Thus we assume that there is a transmission error in the second digit of w. Hence the received word should have been 111111, and we decode this as 11, the first two digits of this word.

3. The syndrome of the given word w is $wA^* = [0\ \ 0\ \ 0\ \ 0]$, and so the received word is a codeword. We decode the received word as 01, its first two digits.

5. The syndrome of the given word w is $wA^* = [0 \ \ 0 \ \ 1 \ \ 1]$, which is not a row of A^*. Thus there are two or more transmission errors in the received word, and so it cannot be decoded reliably.

7. The syndrome of the given word w is $wA^* = [1 \ \ 0 \ \ 0 \ \ 1]$, which is the first row of A^*. Thus we assume that the received word should have been 010110, and we decode it as 01.

9. The check matrix for the given generator matrix A is

$$A^* = \begin{bmatrix} 1 & 0 & 1 & 0 \\ 0 & 1 & 1 & 0 \\ 1 & 1 & 0 & 1 \\ 1 & 0 & 0 & 0 \\ 0 & 1 & 0 & 0 \\ 0 & 0 & 1 & 0 \\ 0 & 0 & 0 & 1 \end{bmatrix}.$$

The syndrome of the given word is $[0 \ \ 0 \ \ 1 \ \ 0]$, which is the sixth row of A^*. Thus we assume that the received word should have been 1010111, and we decode it as its first three digits, 101.

11. Using the check matrix in the solution to Exercise 9, we find that the syndrome is $[1 \ \ 1 \ \ 1 \ \ 0]$, which is not a row of A^*. Hence this received word cannot be decoded reliably.

13. Using the check matrix in the solution to Exercise 9, we find that the syndrome is $[0 \ \ 0 \ \ 0 \ \ 0]$, and so the received word is a codeword. Thus we assume that no transmission errors have occurred, and so we decode the received word as its first three digits, 011.

15. Using the check matrix in the solution to Exercise 9, we find that the syndrome is $[1 \ \ 0 \ \ 1 \ \ 0]$, the first row of A^*. Hence we assume that the received word should have been 0100110, and so we decode the received word as 010.

17. Using the check matrix in the solution to Exercise 9, we find that the syndrome is $[0 \ \ 0 \ \ 0 \ \ 0]$, and so the received word is a codeword. Thus we assume that no transmission errors have occurred, and so we decode the received word as its first three digits, 001.

19. Using the check matrix in the solution to Exercise 9, we find that the syndrome is $[1 \ \ 0 \ \ 1 \ \ 1]$, which is not a row of A^*. Hence this received word cannot be decoded reliably.

21. Using the check matrix in the solution to Exercise 9, we find that the syndrome is $[0 \ \ 1 \ \ 1 \ \ 0]$, the second row of A^*. Hence we assume that the received word should have been 1101100, and so we decode the received word as 110.

23. Using the check matrix in the solution to Exercise 9, we find that the syndrome is [0 0 0 0], and so the received word is a codeword. Thus we assume that no transmission errors have occurred, and so we decode the received word as its first three digits, 110.

25. Using the check matrix in the solution to Exercise 9, we find that the syndrome is [0 0 0 1], the seventh row of A^*. Hence we assume that the received word should have been 1001010, and so we decode the received word as 100.

27. Using the check matrix in the solution to Exercise 9, we find that the syndrome is [1 1 0 1], the third row of A^*. Hence we assume that the received word should have been 0111011, and so we decode the received word as 011.

29. Using the given check matrix, we compute the syndrome of each word in the given list. These syndromes are 010, 101, 011, and 000, respectively. Note that 010 and 011 are the fourth and second rows of the check matrix. Thus the received words are decoded as 11, ??, 11, and 10, respectively.

31. According to Theorem 3.7, a (k,n)-code exists that corrects all single-digit errors if and only if $2^{n-k} - 1 \geq n$. For $k = 8$, we see that

$$2^{11-8} - 1 = 2^3 - 1 = 7 < 11 \qquad \text{but} \qquad 2^{12-8} - 1 = 2^4 - 1 = 15 \geq 12.$$

Hence the minimal value of n for which there exists a $(8, n)$-code that corrects all single-digit errors is 12.

33. We proceed as in the solution to Exercise 31. Note that

$$2^{24-20} - 1 = 2^4 - 1 = 15 < 24 \qquad \text{but} \qquad 2^{25-20} - 1 = 2^5 - 1 = 31 \geq 25.$$

Hence the minimal value of n for which there exists a $(20, n)$-code that corrects all single-digit errors is 25.

35. For $k = 4$ and $n = 7$, the $(4, 7)$-Hamming code has efficiency

$$\frac{k}{n} = \frac{4}{7} > .5.$$

37. For $k = 57$ and $n = 63$, the $(57, 63)$-Hamming code has efficiency

$$\frac{k}{n} = \frac{57}{63} > .9.$$

39. Consider a (k, n)-Hamming code. Let u be a message word that is transmitted with two or more errors as c', and let c be the codeword that results from the transmission of u with no transmission errors. Suppose that check matrix row decoding decodes c' as the codeword c_1. Because c' is decoded as though there is at most one transmission error, $d(c_1, c') \leq 1$. However, $d(c', c) \geq 2$; so $c_1 \neq c$. Because the encoding function is one-to-one, the codewords c_1 and c must differ in their first k digits, and hence c' is decoded incorrectly.

41. Consider the $1 \times k$ message word w in which each digit equals 0. If A is the generator matrix of a (k, n)-block code, then $wA = z$. Hence z is a codeword.

43. Suppose that the check matrix A^* for a (k, n)-block code has two identical rows, say rows i and j. For $r = 1, 2, \ldots, k$, let e_r denote the $1 \times k$ matrix in which all the entries equal 0 except for entry r, which is 1. Then $e_i A^*$ equals the ith row of A^*, and $e_j A^*$ equals the jth row of A^*. Hence

$$(e_i - e_j)A^* = e_i A^* - e_j A^* = O,$$

where O denotes the $1 \times (n - k)$ matrix in which each entry equals 0. It follows from Theorem 3.5(b) that $e_i - e_j$ is a codeword. But also, by Exercise 41 above, the $1 \times n$ matrix z in which each entry equals 0 is also a codeword. Because the Hamming distance between $e_i - e_j$ and z is 2, Theorem 3.4(b) shows that this code cannot correct all single-digit errors.

SUPPLEMENTARY EXERCISES

1. Since $37 - 18 = 19$ is not divisible by 2, $37 \not\equiv 18 \pmod 2$.

3. Since $-7 - 53 = -60$ is divisible by 12, $-7 \equiv 53 \pmod{12}$.

5. In $\mathcal{Z}_{11}$, $[43] + [32] = [75] = [9]$ since 9 is the remainder when 75 is divided by 11.

7. In $\mathcal{Z}_9$, $[5][11] = [55] = [1]$ because 1 is the remainder when 55 is divided by 9.

9. In $\mathcal{Z}_5$, $[22]^7 = [2]^7 = [2^7] = [128] = [3]$ because 3 is the remainder when 128 is divided by 5.

11. We have
$$x^2 + 3y \equiv 4^2 + 3(9) \equiv 16 + 27 \equiv 43 \equiv 10 \pmod{11}.$$

Hence 10 is the remainder when $x^2 + 3y$ is divided by 11.

13. We have $100 = qd + 2$ for some integer q. Hence $98 = qd$, and so

$$198 = 100 + 98 = (qd + 2) + qd = 2qd + 2.$$

Thus the remainder when 198 is divided by d is also 2.

15. Taking $m = 770$ and $n = 1764$ in the Euclidean algorithm leads to the following table.

i	r_i
-1	770
0	1764
1	770
2	224
3	98
4	28
5	14
6	0

Since $r_6 = 0$, $\gcd(770, 1764) = r_5 = 14$.

17. We have $\gcd(-9798, 552) = \gcd(9798, 552)$, and so apply the Euclidean algorithm with $m = 9798$ and $n = 552$.

i	r_i
-1	9798
0	552
1	414
2	138
3	0

Since $r_3 = 0$, $\gcd(-9798, 552) = \gcd(9798, 552) = r_2 = 138$.

19. Taking $m = 770$ and $n = 1764$ in the extended Euclidean algorithm leads to the following table.

i	q_i	r_i	x_i	y_i
-1		770	1	0
0		1764	0	1
1	0	770	1	0
2	2	224	-2	1
3	3	98	7	-3
4	2	28	-16	7
5	3	14	55	-24
6	2	0	-126	55

Since $r_6 = 0$, $\gcd(770, 1764) = r_5 = 14 = 770 \cdot 55 + 1764(-24)$.

21. Taking $m = 9798$ and $n = 552$ in the extended Euclidean algorithm leads to the following table.

i	q_i	r_i	x_i	y_i
-1		9798	1	0
0		552	0	1
1	17	414	1	-17
2	1	138	-1	18
3	3	0	4	-71

Then $\gcd(-9798, 552) = \gcd(9798, 552) = 138$ and

$$138 = 9798(-1) + 552 \cdot 18 = (-9798)1 + 552 \cdot 18.$$

23. Taking $m = 666$ and $n = 1414$ in the extended Euclidean algorithm leads to the following table.

i	q_i	r_i	x_i	y_i
-1		666	1	0
0		1414	0	1
1	0	666	1	0
2	2	82	-2	1
3	8	10	17	-8
4	8	2	-138	65
5	5	0	707	-333

Thus $\gcd(666, 1414) = 2$. Since in (a) the right side is $30 = 15 \cdot 2$, we start with $666(-138) + 1414 \cdot 65 = 2$. Then $666(-138 \cdot 15) + 1414(65 \cdot 15) = 666(-2070) + 1414 \cdot 975 = 15 \cdot 2 = 30$. Thus in (a) we can take $x = -2070$ and $y = 975$. There is no solution to (b) because 2 does not divide 55.

25. We have

$$18^5 = 1889568 = 48450 \cdot 39 + 18,$$
$$10^5 = 100000 = 2564 \cdot 39 + 4,$$
$$6^5 = 7776 = 199 \cdot 39 + 15,$$
$$2^5 = 32 = 0 \cdot 32 + 8.$$

Thus the ciphertext is 18, 4, 15, 32.

27. We start by setting $r_2 = 1$, $p = 18$, and $e = 29$. Since $e \neq 0$, we write $29 = 2 \cdot 14 + 1$; so $Q = 14$ and $R = 1$. Also $p^2 = 18^2 = 324 = 5 \cdot 57 + 39$; so $r_1 = 39$. Since $R = 1$, we write $r_2 p = 1 \cdot 18 = 0 \cdot 57 + 18$, and replace r_2 with 18. Then we set $p = 39$ and $e = 14$. Thus far we have the following table.

Q	R	r_1	r_2	p	e
			1	18	29
14	1	39	18	39	14

Continuing with the algorithm leads to the following.

Q	R	r_1	r_2	p	e
			1	18	29
14	1	39	18	39	14
7	0	39		39	7
3	1	39	18	39	3
1	1	39	18	39	1
0	1	39	18	39	0

Since now $e = 0$, we print $r_2 = 18$.

29. Since $1829 = 31 \cdot 59$, we take $b = (31 - 1)(59 - 1) = 1740$.

31. Since $143 = 11 \cdot 13$, $b = (11 - 1)(13 - 1) = 120$. We must find the smallest positive solution to $11x \equiv 1 \pmod{120}$. Taking $m = 120$ and $n = 11$ in the extended Euclidean algorithm leads to the following table.

i	q_i	r_i	x_i	y_i
-1		120	1	0
0		11	0	1
1	10	10	1	-10
2	1	1	-1	11
3	10	0	11	-120

Thus $120(-1) + 11 \cdot 11 = 1$, and $11 \cdot 11 = 1 + 120 \equiv 1 \pmod{120}$. Since $0 \leq 11 < 120$, we take $D = 11$. Then $6^{11} = 362797056 = 2537042 \cdot 143 + 50$; so $P = 50$.

33. The given block contains exactly eight 1s. Hence the parity check digit is 0.

35. Using formula (3.1), we obtain a probability of

$$C(n, p)p^k(1 - p)^{n-k} = C(10, 4)(.001)^4(.999)^6 = 210(.001)^4(.999)^6 \approx .000000000209.$$

37. The Hamming distance between 1110110111 and 111111111 is 2. This is the minimal Hamming distance between codewords in the given set.

39. Let c_1 denote the codeword in which each digit is 1, and let c_2 denote the codeword other than c_1 that contains the fewest 0s. The minimal Hamming distance between two codewords cannot exceed the minimal Hamming distance between c_1 and c_2, which equals the number of 0s in c_2.

41. An element of $\mathcal{W}_n$ has Hamming distance k from z if it differs from z in precisely k digits. The number of such elements of $\mathcal{W}_n$ is $C(n, k)$.

43. Let A denote the given generator matrix. The codeword corresponding to $w = [1 \ 0 \ 1 \ 0 \ 1]$ is $wA = [1 \ 0 \ 1 \ 0 \ 1 \ 0 \ 0 \ 0 \ 0]$. Note that wA is the sum over $\mathcal{Z}_2$ of the first, third, and fifth rows of A.

45. Multiplying each element of $\mathcal{W}_2 = \{00, 01, 10, 11\}$ by the generator matrix gives the set of codewords, which is $\{0000, 0100, 1011, 1111\}$.

47. The generator matrix for Exercises 43-44 has the form $[I_5|J]$, where

$$J = \begin{bmatrix} 1 & 1 & 1 & 1 \\ 0 & 0 & 1 & 1 \\ 1 & 0 & 1 & 0 \\ 1 & 1 & 0 & 0 \\ 0 & 1 & 0 & 1 \end{bmatrix}.$$

Thus the associated check matrix is

$$A^* = \left[\begin{array}{c} J \\ \hline I_4 \end{array}\right] = \begin{bmatrix} 1 & 1 & 1 & 1 \\ 0 & 0 & 1 & 1 \\ 1 & 0 & 1 & 0 \\ 1 & 1 & 0 & 0 \\ 0 & 1 & 0 & 1 \\ 1 & 0 & 0 & 0 \\ 0 & 1 & 0 & 0 \\ 0 & 0 & 1 & 0 \\ 0 & 0 & 0 & 1 \end{bmatrix}.$$

49. The given generator matrix has the form $[I_3|J]$, where

$$J = \begin{bmatrix} 1 & 0 & 1 & 0 & 1 \\ 0 & 0 & 1 & 1 & 1 \\ 1 & 1 & 0 & 1 & 1 \end{bmatrix}.$$

Its associated check matrix is

$$A^* = \left[\begin{array}{c} J \\ \hline I_5 \end{array}\right] = \begin{bmatrix} 1 & 0 & 1 & 0 & 1 \\ 0 & 0 & 1 & 1 & 1 \\ 1 & 1 & 0 & 1 & 1 \\ 1 & 0 & 0 & 0 & 0 \\ 0 & 1 & 0 & 0 & 0 \\ 0 & 0 & 1 & 0 & 0 \\ 0 & 0 & 0 & 1 & 0 \\ 0 & 0 & 0 & 0 & 1 \end{bmatrix}.$$

51. The given check matrix has the form $\left[\begin{array}{c} J \\ \hline I_3 \end{array}\right]$, where

$$J = \begin{bmatrix} 1 & 0 & 1 \\ 0 & 1 & 1 \end{bmatrix}.$$

The corresponding generator matrix is

$$[I_2|J] = \begin{bmatrix} 1 & 0 & 1 & 0 & 1 \\ 0 & 1 & 0 & 1 & 1 \end{bmatrix}.$$

53. The generator matrix for a (1, 6)-block code is a 1×6 matrix. Since it encodes 1 as 111111, the matrix must be $[1 \ 1 \ 1 \ 1 \ 1 \ 1]$.

55. The check matrix associated with the given generator matrix is

$$
A^* = \left[\begin{array}{c} J \\ \hline I_4 \end{array}\right] = \begin{bmatrix} 1 & 0 & 1 & 1 \\ 0 & 1 & 1 & 1 \\ 1 & 0 & 0 & 1 \\ 1 & 0 & 0 & 0 \\ 0 & 1 & 0 & 0 \\ 0 & 0 & 1 & 0 \\ 0 & 0 & 0 & 1 \end{bmatrix}.
$$

The syndrome of $w = [0\ 1\ 1\ 1\ 1\ 1\ 0]$ is $wA^* = [0\ 0\ 0\ 0]$. Thus, by Theorem 3.5(b), w is a codeword for this code, and so we decode w as its first four digits, 0111.

57. Using the check matrix in the solution to Exercise 55 above, we see that the syndrome of $w = [1\ 0\ 1\ 0\ 1\ 1\ 0]$ is $wA^* = [0\ 1\ 0\ 0]$. Because this is the fifth row of A^*, we assume that the only transmission error is in the fifth digit of w. Hence the correct word is assumed to be 1010010, which we decode as 1010.

59. Using the check matrix in the solution to Exercise 55 above, we see that the syndrome of $w = [1\ 0\ 0\ 1\ 0\ 1\ 1]$ is $wA^* = [0\ 0\ 0\ 0]$. As in Exercise 55, w is a codeword, which is decoded as 1001.

61. The check matrix associated with the given generator matrix A is

$$
A^* = \begin{bmatrix} 1 & 1 & 0 \\ 1 & 0 & 1 \\ 1 & 1 & 1 \\ 0 & 1 & 1 \\ 1 & 0 & 0 \\ 0 & 1 & 0 \\ 0 & 0 & 1 \end{bmatrix}.
$$

  (a) Clearly the rows of A^* include all the nonzero words of length three; so A is a generator matrix for the (4, 7)-Hamming code.
  (b) The syndromes of the three received words are $[1\ 0\ 1]$, $[1\ 0\ 0]$, and $[1\ 0\ 1]$. Since these are the second, fifth, and second rows of A^*, respectively, we assume that the correct words are 0101110, 1110100, and 1101000, which decode as 0101, 1110, and 1101, respectively.
  (c) The codeword corresponding to the message word 001 is 0011001, and the codeword corresponding to the message word 000 is 0000000. Thus the minimal Hamming distance m between distinct codewords satisfies $m \leq 3$. On the other hand, every Hamming code can correct single-digit errors; so $m \geq 3$ by Theorem 3.4(b). It follows that $m = 3$. Hence, by Theorem 3.4(a), this code can detect 2 or fewer errors.
  (d) It follows from (c) above that this code can correct errors in a single digit.

63. Consider a codeword z. By Exercise 41 above, the number of codewords distinct from z whose Hamming distance from z is less than 4 is

$$C(10,1) + C(10,2) + C(10,3) = 10 + 45 + 120 = 175.$$

Because the minimal distance between codewords is 4, none of these 175 words can be codewords. This the maximum number of possible codewords cannot exceed

$$\frac{2^{10}}{176} \approx 5.82.$$

Since the number of codewords is an integer, there can be at most 5 codewords.

65. Let A be the generator matrix for a (k,n)-matrix code. If c and d are codewords, then there are message words u and v in $\mathcal{W}_k$ such that $uA = c$ and $vA = d$. Now $u + v$ is in $\mathcal{W}_k$, and $(u+v)A = uA + vA = c + d$. Thus $c + d$ is a codeword.

67. Consider $a = 4$, $b = 1$, and $c = 3$. Then a divides $b + c$, but a divides neither b nor c.

69. For any integer n greater than 1, we have

$$n^3 + 1 = (n+1)(n^2 - n + 1).$$

Because both $n + 1$ and $n^2 - n + 1$ are integers greater than 1, $n^3 + 1$ is a product of two integers greater than 1 and hence is not prime.

71. If $n = 1$, then $2^n + 3^n = 2 + 3 = 5 = 5^n$. So, for $n = 1$, we have

$$2^n + 3^n \equiv 5^n \pmod{6}.$$

Assume that $2^k + 3^k \equiv 5^k \pmod{6}$ for some positive integer k. Then

$$5^{k+1} \equiv 5(5^k) \equiv 5(2^k + 3^k) \equiv 5(2^k) + 5(3^k) \equiv (2+3)2^k + (2+3)3^k$$

$$\equiv 2^{k+1} + 6(2^{k-1}) + 6(3^{k-1}) + 3^{k+1} \equiv 2^{k+1} + 3^{k+1} \pmod{6}.$$

Hence the congruence holds for $n = k+1$. Therefore, by the principle of mathematical induction, the congruence holds for all positive integers n.

73. If $n = 0$, then $3^{2n+1} + 2(-1)^n = 3^{2(0)+1} + 2(-1)^0 = 3^1 + 2(1) = 5 \equiv 0 \pmod{5}$. Hence the congruence is true for $n = 0$. Assume that it is true for $n = k \geq 0$; then $3^{2k+1} + 2(-1)^k = 5m$ for some integer m. Now

$$\begin{aligned}
3^{2(k+1)+1} + 2(-1)^{k+1} &= 3^{2k+3} - 2(-1)^k \\
&= 9 \cdot 3^{2k+1} - 2(-1)^k \\
&= 9\left[3^{2k+1} + 2(-1)^k\right] - 18(-1)^k - 2(-1)^k \\
&= 9(5m) - 20(-1)^k \\
&= 5\left[9m - 4(-1)^k\right],
\end{aligned}$$

which is divisible by 5. Hence the congruence is true for $k + 1$. Therefore it is true for all nonnegative integers n by the principle of mathematical induction.

Chapter 4

Graphs

4.1 GRAPHS AND THEIR REPRESENTATIONS

1. The points A, B, C, and D are the vertices, and the endpoints of the line segments between two points form the sets which are the edges.

3. The points F, G, and H are the vertices. Since there are no line segments, there are no edges.

5. The elements of $\mathcal{V}$ are the vertices, and the sets $\{B, C\}$, $\{C, A\}$, and $\{B, D\}$ indicate which pairs of points are joined by line segments. A diagram of this graph is given in the answer section of the textbook.

7. There are points representing the vertices G, H, and J, and there are no edges.

9. This is a graph with a single vertex and no edges.

11. One segment does not join two vertices, and so this figure is not a graph.

13. This is not a graph because $\{A, A\}$ cannot be an edge.

15. You, your parents, and grandparents are used as vertices, and vertices that denote persons of the same sex are joined. Two possible graphs are shown below.

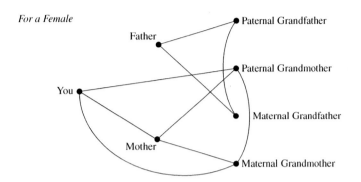

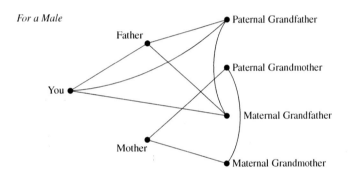

17. Each student is represented by a vertex, and two vertices are joined with an edge if they represent feuding students.

19. Since $\{A, B\}$, $\{A, C\}$, $\{A, D\}$, and $\{A, E\}$ are the only edges on A, A has degree 4 and the vertices B, C, D, and E are adjacent to A. Similarly the vertices adjacent to B are A, C, and F, and so B has degree 3.

21. For (a) there is only one edge on each vertex, and for (b) there are exactly two edges on each vertex.

23. Since $\mathcal{K}_3$ has 3 vertices and an edge connecting any 2 of them, there are $C(3, 2) = 3$ edges. Likewise, $\mathcal{K}_4$ has 4 vertices and an edge connecting any 2 of them. Thus there are $C(4, 2) = 6$ edges. Similarly, $\mathcal{K}_5$ has $C(5, 2) = 10$ edges, and in general, $\mathcal{K}_n$ has

$$C(n, 2) = \frac{n(n - 1)}{2}$$

edges.

25. Since the sum of the degrees (each is 2) of all the vertices is twice the number of edges (which is 10), the number of vertices is 10.

27. Using the procedures described in Examples 4.4 and 4.5, we obtain the adjacency matrix and adjacency list given in the answer section of the textbook.

29. Using the procedures described in Examples 4.4 and 4.5, we obtain the adjacency matrix and adjacency list given in the answer section of the textbok.

31. First we note that there are 4 vertices. Looking at the first row, we see that there are edges from V_1 to each of the other three vertices. This process is continued for each of the other three rows, resulting in the graph given in the answer section of the textbook.

33. We note that there are edges from V_1 to the vertices V_2 and V_3. Continuing in this fashion results in the graph given in the answer section of the textbook.

35. This matrix can be an adjacency matrix since no i, i entry is 1 and whenever the i, j entry is 1, the j, i entry is also 1.

37. This matrix cannot be an adjacency matrix because the $1, 1$ entry is 1.

39. The graphs in Figures 4.5 and 4.6 provide an example to show that the answer is no.

41. The answer is no. For example, consider the following two graphs.

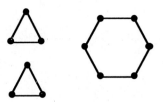

43. (a) The correspondence of A with Z, B with W, C with Y, D with V, E with X, and F with U is an isomorphism.

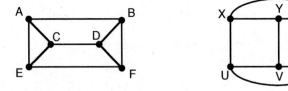

(b) The graphs are not isomorphic since one graph has a vertex of degree 4 and the other does not.

(c) The correspondence of A with W, B with Z, C with Y, D with U, and E with X is an isomorphism.

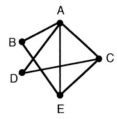
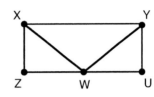

45. We begin by considering the four vertices. Then we look at the possibilities for the number of edges: 0, 1, 2, 3, 4, 5, and 6.

47. We begin by first considering the vertex of degree 3. For the three adjacent vertices we consider the various possible degrees, noting that at least one of these vertices, but not all three, will have degree 1.

49. When n is even, we can join two vertices at a time using $n/2$ edges. In this way, each vertex has degree 1. When n is odd, we can join $n-1$ vertices by joining two vertices at a time using $(n-1)/2$ edges, and then join the unused vertex to one of the others. The total number of edges used is $(n+1)/2$. Furthermore, no fewer edges can be used.

51. To have the smallest number of vertices, the graph should be part of a complete graph $\mathcal{K}_n$ for the smallest positive integer n. To find this n, choose the smallest positive integer n such that
$$m \leq \frac{n(n-1)}{2}.$$

Solving this quadratic inequality yields the answer given in the answer section of the textbook.

53. Since the other 7 people all told Mr. Lewis a different number and each shook hands at most 6 times, these 7 people shook hands 0, 1, 2, 3, 4, 5, and 6 times. The person who shook hands 6 times must be a spouse of the person who shook no hands. Likewise the person who shook hands 5 times must be a spouse of the person who shook 1 other hand, and the person who shook hands 4 times must be a spouse of the person who shook 2 hands. This leaves Mrs. Lewis as the person who shook 3 hands. If we now draw a graph with 8 vertices that include vertices of degrees 0, 1, 2, 3, 4, 5, and 6, it is clear that the remaining vertex must be joined to the vertices of degrees 4, 5, and 6. So Mr. Lewis also shook 3 hands.

4.2 PATHS AND CIRCUITS

1. The multigraph is a graph because there are no loops or parallel edges.

3. The multigraph is not a graph because there are parallel edges.

5. The loops are a and c and there are no parallel edges.

7. The parallel edges are a, b, c, and d and there are no loops.

9. (i) There are many, three of which are given in the answer section of the textbook.

   (ii) There is one simple path from A to D, which is c, and it has length 1.

   (iii) For the paths given in the answer section of the textbook, c is a simple path contained in each.

   (iv) There are only two distinct cycles, namely, those given in the answer section of the textbook.

11. There are many answers. A cycle of length 1 would be a loop, and a cycle of length 2 would be a pair of parallel edges between two vertices.

13. It is possible to find a path between any two distinct vertices.

15. It is not possible to find a path between the vertex in the middle and any other vertex.

17. The multigraph is connected because it is not possible to find two distinct vertices for which there is not a path between them.

19. The multigraph does not have an Euler path because there are more than two vertices with odd degree.

21. The multigraph has an Euler path because there are exactly two vertices of odd degree, and so an Euler path can be found by using the procedure described in Example 4.13.

23. The multigraph does not have an Euler path because there are more than two vertices with odd degree.

25. The multigraph does not have an Euler circuit because not every vertex has even degree.

27. The multigraph does not have an Euler circuit because not every vertex has even degree.

29. The multigraph does not have an Euler circuit because not every vertex has even degree.

31. Represent each of the four different land areas by a vertex, and join two vertices with an edge if there is a bridge between the two corresponding land areas. The resulting multigraph has vertices of degrees 3, 3, 3, and 5. Since each vertex has odd degree, adding a new bridge still leaves some vertices with odd degree, and so there is no Euler circuit.

33. As in Exercise 31, tearing down a bridge is the same as deleting an edge and this still leaves some vertices with odd degree.

35. Successfully tracing the figure is equivalent to finding an Euler circuit. Since each vertex has even degree and the graph is connected, the Euler circuit algorithm can be used to construct an Euler circuit.

37. As in Exercise 35, since each vertex has even degree and the graph is connected, the Euler circuit algorithm can be used to construct an Euler circuit.

39. There are many answers. See Exercise 63 for a description of a general procedure for solving this type of problem.

41. There are many answers. Graphs with Euler circuits can be constructed by using the procedure described in Theorem 4.5. Graphs with Hamiltionian cycles are constructed mainly by trial and error using the definition of a Hamiltionian cycle.

43. These multigraphs are constructed by systematically considering the possibilities as to how a vertex can have degrees 1, 2, and 3.

45. These multigraphs are constructed by systematically considering the possibilities as to how a vertex can have degrees 1, 2, and 3.

47. The isomorphic image of a cycle of length n will again be a cycle of length n.

49. The isomorphic image of a Hamiltonian path will again be a Hamiltonian path.

51. Since one graph has a cycle of length 3 and the other does not, the two graphs are not isomorphic.

53. Since each vertex in $\mathcal{K}_{m,n}$ has degree either m or n, an Euler circuit can be found if and only if both m and n are even.

55. Since each vertex in $\mathcal{K}_n$ has degree $n-1$ and $n-1 > n/2$ when $n > 2$, it follows from Theorem 4.6 that $\mathcal{K}_n$ has a Hamiltionian cycle.

57. Since a vertex has a path of length 0 from itself to itself, the relation is reflexive. Furthermore, because a path from a vertex V to a vertex U is also a path from vertex U to vertex V, the relation is symmetric. Since a path from vertex V to a vertex U and a path from vertex U to a vertex W can be combined to form a path from V to W, the relation is transitive.

59. (a) Since there are paths from vertex A to each of the vertices C, E, and G and to no other vertices, $\{A, C, E, G\}$ is one component of the graph. Similarly $\{B, D, G, H\}$ is another component.

(b) As in (a), we see that $\{I, J, K, L\}$ forms a component.

(c) Since vertex N is adjacent to no other vertices, $\{N\}$ is a component. Then as in (a), we see that $\{M, O, Q\}$ and $\{P, R, S, T\}$ are also components of the graph.

61. In the Hamiltonian cycle, the vertices alternate between men and women, and so the number of men is the same as the number of women.

63. Add an edge between two of the vertices of odd degree and another between the two remaining vertices of odd degree. Then in the resulting multigraph, each vertex has even degree. Thus an Euler circuit can be constructed, and the desired paths can be obtained by deleting the two added edges from this Euler circuit.

65. Let $\mathcal{G}$ be a connected graph with n vertices. We will select a set of $n-1$ distinct edges from $\mathcal{G}$. Pick a vertex V_0 in $\mathcal{G}$. If $n = 1$, then clearly $\mathcal{G}$ has at least $n - 1 = 0$ edges. So assume that $n \geq 2$. Let U be a vertex in $\mathcal{G}$ other than V_0. Because $\mathcal{G}$ is connected, there is a simple path from V_0 to U. Let e_1 denote the edge of this path incident with V_0, and let V_1 be the vertex on e_1 other than V_0. Assume that, for some k, $(1 \leq k < n-1)$ vertices $V_0, V_1, \ldots, V_k$ and edges $e_1, e_2, \ldots, e_k$ have been chosen so that

> (i) $V_0, V_1, \ldots, V_k$ are distinct, and
>
> (ii) for $i = 1, 2, \ldots, k$, edge e_i joins V_i to some vertex $V_0, V_1, \ldots, V_{i-1}$.

Let W be a vertex of $\mathcal{G}$ other than $V_0, V_1, \ldots, V_k$. Since $\mathcal{G}$ is connected, there is a path from V_0 to W. Some edge e_{k+1} of this path must join a vertex V_j $(0 \leq j \leq k)$ to a vertex V_{k+1} that is not in $\{V_0, V_1, \ldots, V_k\}$. Continuing in this manner, we obtain by mathematical induction a set of edges $\mathcal{S} = \{e_1, e_2, \ldots, e_{n-1}\}$ in $\mathcal{G}$. If $i < j$, then e_i joins V_i to one of $V_0, V_1, \ldots, V_{i-1}$. Since the vertices $V_0, V_1, \ldots, V_n$ are distinct, V_j is not incident with e_i. However, V_j is incident with e_j, and so $e_i \neq e_j$. Thus $\mathcal{S}$ is a set of $n - 1$ edges, and so $\mathcal{G}$ contains at least $n - 1$ edges.

4.3   SHORTEST PATHS AND DISTANCE

1. As in Example 4.17, we have the following labels.

Vertex	Label	Predecessor
S	0	none
A, D	1	S, S
B, G	2	D, D
E, H	3	G, G
C, F, I	4	E, E, H
T	5	F

So the distance from S to T is 5. A shortest path from S to T is S, D, G, E, F, T. This path is formed in reverse order by looking at T, then its predecessor F, then the predecessor E of F, and so forth until S is reached.

3. As in Example 4.17, we have the following labels.

Vertex	Label	Predecessor
S	0	none
A, D	1	S, S
E, F	2	D, A
G, I, J	3	F, E, F
B, H, M, N	4	G, I, I, J
C, L, Q	5	B, M, N
O, P, R	6	C, L, Q
T, U	7	O, O

So the distance from S to T is 7, and a shortest path of length 7 from S to T is formed as in Exercise 1. This shortest path is S, A, F, G, B, C, O, T.

5. Dijkstra's Algorithm is applied as in Example 4.20. In the initial assignment, S has label $0(-)$, A has $\infty(S)$, B has $\infty(S)$, C has $3(S)$, D has $\infty(S)$, E has $\infty(S)$, F has $\infty(S)$, G has $\infty(S)$, H has $5(S)$, and I has $\infty(S)$. We start with $\mathcal{P} = \{S\}$.

Vertex Included in $\mathcal{P}$	Changes in Labels
C	D has $5(C)$, E has $4(C)$
E	F has $6(E)$
D	A has $10(D)$
H	I has $6(H)$
F	G has $7(F)$
I	B has $10(I)$
G	A has $8(G)$, B has $9(G)$
A	none
B	none

So the final assignment of labels is: S has $0(-)$, A has $8(G)$, B has $9(G)$, C has $3(S)$, D has $5(C)$, E has $4(C)$, F has $6(E)$, G has $7(F)$, H has $5(S)$, and I has $6(H)$. The distances from S to A, B, C, D, E, F, G, H, and I, given by the labels, are 8, 9, 3, 5, 4, 6, 7, 5, and 6, respectively. A shortest path from S to A is found by looking at the predecessors and backtracking through them. Finding a shortest path from S to B is done similarly.

7. Proceeding as in Exercise 5, we find the final assignment of labels to be: S has $0(-)$, A has $8(G)$, B has $6(H)$, C has $3(S)$, D has $5(C)$, E has $2(S)$, F has $3(H)$, G has $6(D)$, and H has $1(S)$. The distance from S to each of the other vertices is found by looking at the labels. Shortest paths from S to A and B are found by backtracking through the predecessors as described in the solution to Exercise 5.

9. A shortest path from S to T that goes through the vertex A must necessarily be made up of a shortest path from S to A and a shortest path from A to T. Thus the strategy is to find a shortest path from S to A and a shortest path from S to T and join these two paths together. The most efficient way to do this is to apply Dijkstra's algorithm starting at A to find a shortest path from A to every other vertex. The final label assignments are: A has $0(-)$, B has $4(A)$, C has $2(A)$, D has $9(C)$, E has $6(F)$, F has $4(K)$, G has $1(A)$, H has $4(C)$, I has $6(H)$, J has $8(K)$, K has $3(L)$, L has $2(G)$, M has $5(L)$, S has $9(E)$, and T has $8(M)$. Thus one shortest path from S to T through A is $S, E, F, K, L, G, A, G, L, M, T$.

11. We proceed as in Exercise 9. In finding a shortest path from A to the other vertices, the final label assignments are: A has $0(-)$, B has $3(F)$, C has $1(A)$, D has $3(C)$, E has $4(D)$, F has $2(A)$, G has $3(A)$, H has $1(A)$, I has $7(T)$, J has $2(H)$, K has $3(J)$, L has $6(K)$, S has $3(F)$, and T has $5(E)$. Thus a shortest path from S to T through A is S, F, A, C, D, E, T.

13. The adjacency matrix is

$$A = \begin{bmatrix} 0 & 1 & 1 & 1 \\ 1 & 0 & 1 & 1 \\ 1 & 1 & 0 & 1 \\ 1 & 1 & 1 & 0 \end{bmatrix}.$$

By examining the 1,2 entry of A, we see that there is exactly one path of length 1 from V_1 to V_2. Likewise we examine the 2,3 entry of A to see that there is exactly one path of length 1 from V_2 to V_3. Next we compute A^2 and look at the 1,2 entry, which is a 2. This indicates that there are two paths of length 2 from V_1 to V_2. The 2,3 entry of A^2 being 2 indicates there are two paths of length 2 from V_2 to V_3. Similarly the 1,2 entry of A^3 indicates there are 7 paths of length 3 from V_1 to V_2, and the 2,3 entry indicates that there are 7 paths of length 3 from V_2 to V_3. The 1,2 and 2,3 entries of A^4 give the number of paths from V_1 to V_2 and from V_2 to V_3, respectively.

15. The adjacency matrix is

$$A = \begin{bmatrix} 0 & 1 & 1 & 1 & 1 \\ 1 & 0 & 0 & 1 & 1 \\ 1 & 0 & 0 & 0 & 1 \\ 1 & 1 & 0 & 0 & 1 \\ 1 & 1 & 1 & 1 & 0 \end{bmatrix}.$$

As in Exercise 13, the 1,1 and 4,3 entries of A, A^2, A^3, and A^4 give the number of paths of lengths 1, 2, 3, and 4 from V_1 to V_1 and V_4 to V_3, respectively.

17. Since the entries of A, A^2, and A^3 give the respective number of paths of lengths 1, 2, and 3 between all pairs of vertices, the entries of the sum of A, A^2, and A^3 give the number of paths with length at most 3 between any two vertices.

19. We shall prove Theorem 4.7 by mathematical induction on m. When $m = 1$, the definition for the adjacency matrix A gives the result. We assume the theorem is true for r and prove it for $r + 1$. Let b_{ik} be the i, k entry of A^r, which gives the number of paths of length r from V_i to V_k. For a path of length $r + 1$ from V_i to V_j, there is a vertex V_k such that this path is composed of a path of length r from V_i to V_k and an edge from V_k to V_j. So there are $b_{ik}a_{kj}$ paths of length $r + 1$ from V_i to V_j with V_k as the vertex immediately preceding V_j. Then the total number of paths of length $r + 1$ from V_i to V_k is $b_{11}a_{1j} + \cdots + b_{in}a_{nj}$, which is the i, j entry of $A^r A = A^{r+1}$. This establishes the result for $r + 1$ and so completes the induction argument.

21. Suppose there is a shortest path $\mathcal{B}$ from S to U that contains a vertex other than U that is not in $\mathcal{P}$. Thus there is a vertex $W \neq U$ such that W is not in $\mathcal{P}$ and all the vertices on $\mathcal{B}$ before W are in $\mathcal{P}$. Because $\mathcal{B}$ is a shortest path from S to U, the part of the path $\mathcal{B}$ between S and W is then a shortest path $\mathcal{C}$ from S to W with the restriction that W is the only vertex of the path $\mathcal{C}$ not in $\mathcal{P}$. Then by condition (ii), $\mathcal{C}$ has length $L(W)$. Since all the weights are positive and W is on the path $\mathcal{B}$, $L(W) < L(V)$, which is a contradiction.

75

23. Each vertex in $\mathcal{P}$ satisfies condition (i), and Exercise 22 shows that U satisfies condition (i). For convenience, in the rest of this exercise we shall use the phrase "$\mathcal{P}'$-path from S to V" to mean a path from S to V such that V is the only vertex of the path not in $\mathcal{P}'$." Let V be a vertex not in $\mathcal{P}'$.

Assume first that there is a shortest $\mathcal{P}'$-path from S to V that does not include U. By condition (ii) in Exercise 21, $L(V)$ is the length of a shortest $\mathcal{P}'$-path from S to V. Thus, even if there is an edge joining U and V, the shortest path from S to U followed by the edge from U to V (which forms a $\mathcal{P}'$-path from S to V) cannot have length less than $L(V)$. Therefore the length of a shortest $\mathcal{P}'$-path from S to V is $L(V)$, which cannot be more than $L(U) + W(U, V)$.

Now suppose there is a shortest $\mathcal{P}'$-path $\mathcal{B}$ from S to V that includes U. Let W denote the vertex that immediately precedes V on $\mathcal{B}$. If $W \neq U$, there is a path $\mathcal{C}$ from S to W using only vertices in $\mathcal{P}$. Following $\mathcal{C}$ by the edge from W to V produces a shortest $\mathcal{P}'$-path from S to V that does not include U. Therefore the conclusion follows as in the preceding paragraph. Otherwise, if $W = U$, then $\mathcal{B}$ includes the edge joining U and V. Since the part of $\mathcal{B}$ from S to U is a shortest path from S to U having all its vertices in $\mathcal{P}$, the length of this part of $\mathcal{B}$ is $L(U)$. Hence the length of $\mathcal{B}$, a shortest $\mathcal{P}'$-path from S to V, is $L(U) + W(U, V)$. Thus, in either case, the minimum of $L(V)$ and $L(U) + W(U, V)$ is the length of a shortest path from S to V.

4.4   COLORING A GRAPH

1. Because there are three vertices adjacent to each other, at least three colors must be used. There are many solutions using three colors, one is given in the answer section of the textbook.

3. Because there are three vertices adjacent to each other, at least three colors must be used. There are many solutions using three colors, one is given in the answer section of the textbook.

5. Because there are three vertices adjacent to each other, at least three colors must be used. There are many solutions using three colors, one is given in the answer section of the textbook.

7. Because there are two vertices adjacent to each other, at least two colors must be used. There are many solutions using two colors, one is given in the answer section of the textbook.

9. When a graph has chromatic number 1, there can be no pair of adjacent vertices. Hence there are no edges in the graph.

11. (a) There are many answers, one of which is given in the answer section of the textbook.

(b)   There are many answers, one of which is given in the answer section of the textbook.

13.   There are many answers, one of which is given in the answer section of the textbook.

15.   Applying the algorithm in Exercise 13 yields a coloring using two colors. One possible answer is given in the answer section of the textbook.

17.   Applying the algorithm in Exercise 13 yields a coloring using two colors. One possible answer is given in the answer section of the textbook.

19.   Since vertices [2, 3], [2, 4], [1, 3], [1, 4], and [3, 4] are all adjacent to each other, at least 5 colors are needed. However, since [1, 2] and [3, 4] are not adjacent, they can be assigned the same color and so the chromatic number is 5.

21.   Since there are n choices of colors for each of the n vertices, the multiplication principle from Section 1.2 says there are n^n ways to assign n colors to the n vertices of the graph.

23.   If we let each region of the map be a vertex of a graph and join vertices representing regions with a common boundary other than a point, then we have a graph with at least three adjacent vertices. Thus at least three colors are needed. There are many solutions using three colors, one is given in the answer section of the textbook.

25.   There are many solutions using four colors, one is given in the answer section of the textbook.

27.   If we let each committee be a vertex of a graph and join the vertices with an edge when the corresponding two committees cannot meet, this results in a graph with chromatic number 3. Thus three separate meeting times are needed, with finance and agriculture being able to meet at the same time and likewise for labor and budget.

29.   If we let each animal be a vertex of a graph and join two vertices with an edge when one of the corresponding animals is a predator of the other, this results in a graph with a cycle of length 5, namely a, b, d, f, j, a. Thus at least three colors are needed. Since each of the remaining vertices are adjacent to at most two of the vertices in the cycle, they can be colored with one of the three colors needed for the cycle. Thus three is the minimum number of locations needed by the zookeeper.

31.   Suppose f is an isomorphism of the graph $\mathcal{G}_1$ with the graph $\mathcal{G}_2$ and that $\mathcal{G}_1$ can be colored with three colors. We assign to each vertex in $\mathcal{G}_2$ the color assigned to the corresponding vertex (with respect to f) in $\mathcal{G}_1$. Since adjacency is a graph isomorphism invariant, $\mathcal{G}_2$ will have chromatic color 3.

33.   When a graph with n vertices has chromatic color n, each vertex is assigned a different color. We will show there is an edge between any two vertices. Suppose there is no edge joining vertices V and W. Then V and W can be assigned the same color, and the other vertices can be colored using $n - 2$ other colors. This colors the graph with $n - 1$ colors, which is a contradiction. Thus the graph is $\mathcal{K}_n$ and has $n(n-1)/2$ edges.

35. There are six vertices joined in pairs by edges, each having been assigned the color red or blue. We select a vertex and label it V_1. There are edges from V_1 to each of the other five vertices. Of these five edges, at least three of them must be of the same color, say red. For these three red edges, we label the other vertices as V_2, V_3, and V_4. If any one of the edges $\{V_2, V_3\}$, $\{V_2, V_4\}$, and $\{V_3, V_4\}$ is red, then there is a cycle of length three with all red edges. Otherwise, these edges are all colored blue and form a cycle of length three with all its edges the same color.

37. Suppose $\mathcal{G}$ is a graph with chromatic number k that has the property stated in the first sentence of the exercise. Now suppose V is a vertex of $\mathcal{G}$ that has degree less than $k - 1$. Then V is adjacent to at most $k - 2$ vertices, and the graph formed by V and the vertices adjacent to V can be colored using at most $k - 1$ colors. Therefore deleting V and the edges incident on V will not decrease the number of colors needed to color $\mathcal{G}$, which is a contradiction.

4.5 DIRECTED GRAPHS AND MULTIGRAPHS

1. The points A, B, C, and D are the vertices, and the endpoints of the directed line segments between two points form the ordered pairs which are the directed edges.

3. The points A, B, C, and D are the vertices, and the endpoints of the directed line segments between two points form the ordered pairs which are the directed edges.

5. The elements of $\mathcal{V}$ are the vertices, and the ordered pairs indicate which pairs of vertices are joined by directed line segments. A diagram of this directed graph is given in the answer section of the textbook.

7. The elements of $\mathcal{V}$ are the vertices, and the ordered pairs indicate which pairs of vertices are joined by directed line segments. A diagram of this directed graph is given in the answer section of the textbook.

9. First we note that there are 4 vertices. Looking at the first row, we see that there are directed edges from V_1 to V_2 and V_4. This process is continued for each of the other three rows, resulting in the directed graph given in the answer section of the textbook.

11. First we note that there are 4 vertices. Looking at the first row, we see that there are directed edges from V_1 to V_2, V_3, and V_4. This process is continued for each of the other three rows, resulting in the directed graph given in the answer section of the textbook.

13. We note that there are directed edges from B and C to A and also directed edges from A to B and D. So the indegree of A is 2, and the outdegree of A is 2.

15. We note that there are directed edges from B, C, D, and E to A and no directed edges from A to any other vertex. So the indegree of A is 4, and the outdegree of A is 0.

17. (i) A simple directed path is a path without repeated vertices. The length of a simple directed path is the number of directed edges in the directed path. The answer is given in the answer section of the textbook.

    (ii) A directed cycle is a directed path of positive length from some vertex to itself with no other vertices being repeated, and the length is the number of directed edges in the directed cycle. The answer is given in the answer section of the textbook.

19. Using the procedures described in Examples 4.34 and 4.35, we obtain the adjacency matrix and adjacency list as given in the answer section of the textbook.

21. Using the procedures described in Examples 4.34 and 4.35, we obtain the adjacency matrix and adjacency list as given in the answer section of the textbook.

23. The vertices are $1, 2, \ldots, 10$. Since 1 divides each of $2, \ldots, 10$ there is a directed edge from 1 to each of these vertices. Since 2 divides 4, 6, 8, and 10, there is a directed edge from 2 to each of 4, 6, 8, and 10. This process is continued, resulting in the directed graph given in the answer section of the textbook.

25. We begin with the two vertices. Then we look at the possibilities for the directed edges between two vertices, resulting in the answer as given in the answer section of the textbook.

27. The vertices are 3, 5, 8, 10, 15, and 24. Since 3 divides each of 3, 15, and 24, there is a directed edge from 3 to each of these vertices. Since 5 divides 5, 10, and 15, there is a directed edge from 5 to each of these vertices. This process is continued, resulting in the directed graph given in the answer section of the textbook.

29. A relation on a set is reflexive when $x \, R \, x$ for every element x of the set. For a directed multigraph, this requires a directed edge from each vertex to itself.

31. A relation on a set is antisymmetric whenever $x \, R \, y$ and $y \, R \, x$ imply $x = y$. In the case of a directed multigraph, when there is a directed edge from A to B and also one from B to A, then we must have $A = B$.

33. Reverse the direction of the directed line segments of the directed graph in Exercise 32, resulting in the directed graph given in the answer section of the textbook. As a consequence, (A, B) is a directed edge in the directed graph of Exercise 32 if and only if (B, A) is a directed edge in the directed graph of Exercise 33.

35. By inspection, we can find a directed path from A to each of the vertices B, C, and D, and from B to each of A, C, and D, and from C to each of A, B, and D, and finally from D to each of A, B, and C. Thus the directed graph in Exercise 17 is strongly connected.

37. There are many answers, one of which is given in the answer section of the textbook.

39. An acceptable assignment is not possible because the removal of the edge in the upper left portion of the graph results in a vertex with no edges, so that the resulting graph is not connected. In other words, assigning a direction to the edge in the upper left portion of the graph will make it impossible to have a directed path both to and from the vertex in the upper left corner of the graph.

41. Since the removal of any one edge still leaves a graph that is connected, the graph is strongly connected. One possible assignment of directions to the edges that produces a strongly connected graph is given in the answer section of the textbook.

43. Because a directed Hamiltonian cycle contains every vertex of the directed graph, it is possible to find a directed path from any one vertex to another by following the directed Hamiltonian cycle.

45. Whenever the directions on the directed multigraph are ignored, the resulting graph is connected. Furthermore, the indegree of each vertex is the same as the outdgree. Thus by Theorem 4.13, a directed Euler circuit may be constructed by using the modification suggested in the paragraph following Theorem 4.13. One example of a directed Euler circuit is given in the answer section of the textbook.

47. Whenever the directions on the directed multigraph are ignored, the resulting graph is connected. Furthermore, the indegree of each vertex is the same as the outdgree except for two distinct vertices where the outdegree of one exceeds its indegree by 1 and the indegree of the second exceeds its outdegree by 1. Thus by Theorem 4.13, a directed Euler path may be constructed by following the modification suggested in the paragraph following Theorem 4.13. One example of a directed Euler path is given in the answer section of the textbook.

49. There is neither since there is a vertex with indegree 3 and outdegree 1, and so the conditions of Theorem 4.13 are not satisfied.

51. In a tournament there are $n(n-1)/2$ directed edges, and so, by Theorem 4.10, the sum of the outdegrees is $n(n-1)/2$.

53. Because the indegree of B is 0, a directed Hamiltonian path must begin at B. Since the outdegree of D is 0, a directed Hamiltonian path must end at D. Looking at the possibilities, we see that there is only one directed Hamiltonian path, which is given in the answer section of the textbook.

55. We represent the situation by a directed graph in which the vertices are labeled with the five desserts and a directed edge from one dessert to another indicates a preference for the first dessert over the second. The resulting directed graph is a tournament. A ranking is equivalent to a directed Hamiltonian path. Since cookies has indegree 0 and pudding has outdegree 0, a directed Hamiltonian path must begin with cookies and end with pudding. Looking at the possibilities, there is only one directed Hamiltonian path and hence only one ranking as given in the answer section of the textbook.

57. There are many answers since each of the vertices B and C has maximum outdegree, namely 3. One possible answer is given in the answer section of the textbook.

59. We represent the situation by a directed graph. The vertices are labeled with the four teams, and there is a directed edge from one team to another if the first team beat the second one. The resulting directed graph is a tournament. A ranking is equivalent to a directed Hamiltonian path. Since Bears has indegree 0 and Lions has outdegree 0, a directed Hamiltonian path must begin with Bears and end with Lions. Looking at the possibilities there is only one directed Hamiltonian path and hence only one ranking as given in the answer section of the textbook.

61. To obtain the algorithm, change the while statement in step 2.2 of the breadth-first search algorithm in Section 4.3 to read as follows: **while** $\mathcal{L}$ contains a vertex V with label $k-1$ such that there is a vertex W not in $\mathcal{L}$ for which there is a directed edge (V, W).

63. Apply the algorithm in Exercise 61 as follows.

Vertex	Label	Predecessor
S	0	none
B	1	S
G	2	B
M, N	3	G
F, H, L, V	4	M, N, M, M
C	5	H
A, D	6	C
E, I	7	D
O, Q	8	C, I
J, R, X	9	Q, Q, O
K, U, Y	10	J, R, X
T	11	K

Starting with T and taking the predecessors in reverse order yields the shortest directed path given in the answer section of the textbook.

65. We use the same procedure as in Exercise 63 to yield the shortest directed path given in the answer section of the textbook.

67. We use the algorithm in Exercise 66. In the initial assignment, S has label $0(-)$, A has $\infty(S)$, B has $\infty(S)$, C has $5(S)$, D has $3(S)$, E has $\infty(S)$, F has $2(S)$, and G has $\infty(S)$ and $\mathcal{P} = \{S\}$. Applying the algorithm produces the following.

Vertex Included in $\mathcal{P}$	Changes in Labels
F	E has $5(F)$, G has $4(F)$
D	C has $4(D)$
C	A has $6(C)$
G	B has $10(G)$, A has $5(G)$
E	none
A	none
B	none

The final label assignments are as follows: S has $0(-)$, A has $5(G)$, B has $10(G)$, C has $4(D)$, D has $3(S)$, E has $5(F)$, F has $2(S)$, and G has $4(F)$. From these we are able to determine the distances from S to all the other vertices and a shortest directed path from S to A as given in the answer section of the textbook.

69. We proceed as in Exercise 67 to determine the final label assignments: S has $0(-)$, A has $7(H)$, B has $11(I)$, C has $5(S)$, D has $6(C)$, E has $14(D)$, F has $2(S)$, G has $5(F)$, H has $6(G)$, and I has $8(H)$. From these, we are able to determine the distances from S to all the other vertices and a shortest directed path from S to A as given in the answer section of the textbook.

71. We use the same argument as in Exercise 19 of Section 4.3 with references to a path replaced by references to a directed path.

73. The adjacency matrix for the directed graph is

$$A = \begin{bmatrix} 0 & 1 & 1 & 0 \\ 1 & 0 & 1 & 1 \\ 0 & 0 & 0 & 1 \\ 1 & 0 & 1 & 0 \end{bmatrix}.$$

Using Exercise 71, we look at the 1,1 and 4,1 entries of A, A^2, A^3, and A^4 to find the number of directed paths of lengths 1, 2, 3, and 4 from V_1 to V_4 and V_4 to V_1, respectively.

75. An isomorphism from one directed graph to another directed graph preserves not only adjacency between vertices but also the direction of the line segment. There are many answers, and two are given in the answer section of the textbook.

77. **(a)** One directed graph has a vertex with outdegree 3 and the other does not.
(b) One directed graph has a vertex with indegree 0 and the other does not.

79. We recall that each 1 in row i corresponds to a directed edge from the vertex V_i to some other vertex. Thus, the number of 1s in row i is the number of directed edges leaving V_i, which is the outdegree of V_i. A similar argument shows that the sum of the entries in column j equals the indegree of the vertex V_j.

81. The proof of Theorem 4.13 can be established by modifying Theorem 4.5. References to a path in Theorem 4.5 should be changed to references to a directed path, and references to a degree should be change to the appropriate indegree or outdegree.

83. Let A be a vertex of maximum score (outdegree) in a tournament. Suppose there is a vertex W for which there is no directed path of length 1 or 2 from A to W. If A has outdegree k, then there are k vertices $V_1, V_2, \ldots, V_k$ such that there is a directed edge from A to each of them. Consider the directed edges between each V_i and W. Since there is no directed path of length 2 from A to W, each of the directed edges between V_i and W must be directed from W to V_i. Because there is no directed path of length 1 from A to W, the edge between W and A must be directed from W to A. Thus W has outdegree at least $k + 1$, which is a contradiction.

SUPPLEMENTARY EXERCISES

1. One way to construct the complement of a graph $\mathcal{G}$ is to consider the complete graph on the vertices of $\mathcal{G}$ and then delete those edges that correspond to the edges of $\mathcal{G}$. A diagram for the complement of the graph in Exercise 1 is given in the answer section of the textbook.

3. Line segments are drawn between two vertices when they are adjacent. A diagram of the graph is given in the answer section of the textbook.

5. When two graphs are isormorphic, they need to have the same number of vertices of each degree. Here the first graph has a vertex of degree 2 and the second does not. So the two graphs are not isomorphic.

7. We first note that there are five vertices. The adjacency list indicates that vertex V_1 is adjacent to vertices V_2, V_3, and V_5, and so line sements (edges) are drawn between V_1 and V_2, between V_1 and V_3, and between V_1 and V_5. This process is repeated for each of the other 4 vertices, resulting in the graph given in the answer section of the textbook.

9. Represent the front, the back, and each room by a vertex and connect two vertices with an edge when there is a door between the two corresponding areas. This is a graph with more than two vertices of odd degree. Since a successful walk through the house is equivalent to an Euler path, it is not possible to find such a walk.

11. Represent each of the four different land areas by a vertex, and join two vertices with an edge if there is a bridge between the two corresponding land areas. The resulting multigraph has vertices of degrees 3, 3, 3, and 5. Tearing down a bridge leaves a multigraph with two vertices of even degree and two with odd degree. Building a bridge between the two land masses now represented by vertices of odd degree gives a multigraph where every vertex has even degree, and so an Euler circuit can be constructed. This Euler circuit corresponds to a successful walk.

13. The multigraph has an Euler circuit because every vertex has even degree. There are many Euler circuits, one of which is given in the answer section of the textbook.

15. Suppose the graphs $\mathcal{G}_1$ and $\mathcal{G}_2$ are isomorphic and that $\mathcal{G}_1$ is connected. Then for any two vertices A_2 and B_2 in $\mathcal{G}_2$, we look at the corresponding vertices A_1 and B_1 in $\mathcal{G}_1$. Because $\mathcal{G}_1$ is connected, there is a path between A_1 and B_1. The isomorphic image of this path is then a path between A_2 and B_2, and so $\mathcal{G}_2$ is connected.

17. As in Example 4.17, we have the following labels.

Vertex	Label	Predecessor
S	0	none
A, C, D	1	S, S, S
G, H	2	C, D
E, K, L	3	H, H, H
B, F, I	4	E, E, E
J, M	5	F, I
T	6	J

So the distance from S to T is 6. A shortest path is given in the answer section of the textbook.

19. Dijkstra's algorithm is applied as in Example 4.20. In the initial assignment S has label $0(-)$, C has $2(S)$, E has $4(S)$, and the rest have $\infty(S)$. Initially $\mathcal{P} = \{S\}$.

Vertex Included in $\mathcal{P}$	Changes in Labels
C	D has $7(C), F$ has $5(C), E$ has $3(C)$
E	F has $4(E), J$ has $5(E)$
F	G has $6(F)$
J	K has $9(J)$
G	H has $9(G)$
D	H has $8(D)$
H	A has $11(H)$
A	I has $12(A)$
K	B has $13(K)$
I	L has $13(I)$
B	none

The final label on a vertex gives its distance from S. These are listed in the answer section in the textbook. Taking the predecessors in reverse order gives the shortest path in the answer section in the textbook.

21. The adjacency matrix is

$$A = \begin{bmatrix} 0 & 0 & 1 & 0 \\ 0 & 0 & 1 & 1 \\ 1 & 1 & 0 & 1 \\ 0 & 1 & 1 & 0 \end{bmatrix}.$$

As in Exercise 13 of section 4.3, we look at the 1,2 and 1,4 entries of A, A^2, A^3, and A^4 to find the number of paths of lengths 1, 2, 3, and 4 from V_1 to V_2 and V_1 to V_4, respectively.

23. Since there are regions that have boundaries with more than a point in common, at least two colors are needed. Two colors are sufficient, as shown in the figure in the answer section of the textbook.

25. Since there are regions that have boundaries with more than a point in common, at least two colors are needed. By using the following recursive procedure, we see that two colors can be used to colored the map. Draw the first line segment across a square and color the two regions using two colors. For each successive line segment drawn across the square, on one side of the newly drawn line reverse the existing colors in the regions.

27. We let each stack be a vertex and join two vertices with an edge when the two stacks will be in use at the same time. This gives a graph with the following adjacency list.

$$
\begin{aligned}
S_1: &\quad S_4, S_5, S_7, S_9, S_{10} \\
S_2: &\quad S_5, S_6, S_8, S_{10} \\
S_3: &\quad S_6, S_7, S_9 \\
S_4: &\quad S_1, S_7, S_8, S_{10} \\
S_5: &\quad S_1, S_2, S_8, S_9 \\
S_6: &\quad S_2, S_3, S_9, S_{10} \\
S_7: &\quad S_1, S_3, S_4, S_{10} \\
S_8: &\quad S_2, S_4, S_5 \\
S_9: &\quad S_1, S_3, S_5, S_6 \\
S_{10}: &\quad S_1, S_2, S_4, S_6, S_7
\end{aligned}
$$

From the adjacency list we see that S_1, S_4, S_7, and S_{10} are all adjacent to each other, and so at least 4 colors are needed. Thus assigning S_2 and S_3 the same color as S_4, S_5 and S_6 the same color as S_7, S_8 the same color as S_1, and S_9 the same color as S_{10} shows the graph to have chromatic color 4. Hence four stacks are needed.

29. First we note that there are 4 vertices. Looking at the first row, we see that there are directed edges from V_1 to each of V_3 and V_4. This process is continued for each of the other three rows, resulting in the graph given in the answer section of the textbook.

31. We begin with 6 vertices and consider what it means to have outdegrees of 5 and 0 and an indegree of 5. Considering various possibilities leads to the graph in the answer section of the textbook.

33. We construct a directed multigraph as follows. Use 000, 001, 010, 100, 110, 101, 011, 111 as vertices. Next construct two directed edges from each vertex in the manner described in Example 4.38. This will be a directed multigraph in which a directed Euler circuit exists. Construct a directed Euler circuit, and select the first digit of the label of each directed edge in the directed Euler circuit. This yields a sequence of 16 numbers such that every sequence of 4 consecutive entries is different. One possible answer is given in the answer section of the textbook.

35. Suppose U and V are vertices in a transitive tournament with the same score (outdegree), say k. Then there are vertices $U_1, U_2, \ldots, U_k$ having a directed edge from U to each of them. Likewise there are vertices $V_1, V_2, \ldots, V_k$ and a directed edge from V to each of them. We know there is a directed edge between U and V, and so let us say that it is directed from U to V. Thus V is U_i for some i. So we have a directed edge from U to $V = U_i$ and a directed edge from V to each V_j. Since the tournament is transitive, there is a directed edge from U to each V_j, and so the vertices $U_1, U_2, \ldots, U_k$ are the same as the vertices $V_1, V_2, \ldots, V_k$. Thus V is some V_i, which is a contradiction.

37. The adjacency matrix for the directed graph is

$$A = \begin{bmatrix} 0 & 1 & 0 & 1 & 0 \\ 1 & 0 & 1 & 0 & 0 \\ 1 & 0 & 0 & 0 & 1 \\ 0 & 0 & 1 & 0 & 1 \\ 0 & 0 & 0 & 1 & 0 \end{bmatrix}.$$

As in Exercise 73 of Section 4.5, we look at the 1,4 and 2,5 entries of A, A^2, A^3, and A^4 to find the number of directed paths of lengths 1, 2, 3, and 4 from V_1 to V_4 and V_2 to V_5, respectively.

39. A relation on a set is reflexive when $x \, R \, x$ for every element x of the set. Thus, in the case of a directed multigraph, there must be a directed edge from each vertex to itself. A relation on a set is symmetric whenever $x \, R \, y$ implies $y \, R \, x$. In the case of a directed multigraph, this means that whenever there is a directed edge from A to B, then there is also one from B to A. Finally, a relation on a set is transitive whenever $x \, R \, y$ and $y \, R \, z$ imply $x \, R \, z$. In the case of a directed multigraph, this occurs if the existence of directed edges from A to B and from B to C implies that there is a directed edge from A to C.

Chapter 5

Trees

5.1 PROPERTIES OF TREES

1. The graph is a tree since it is connected and has no cycles.

3. The graph is not a tree since it is not connected.

5. The graph is not a tree since it is not connected.

7. The graph is a tree since it is connected and has no cycles.

9. There are 16 vertices since the number of vertices in a tree exceeds the number of edges by one.

11. A computer telecommunication network can be modeled by a graph where the towns are represented by vertices and the communication lines by edges. Since the network is to have as few lines as possible, there should be no cycles in the graph. Secondly, since there is to be communication between any two towns, the graph should be connected. Hence, the graph representing the desired network should be a tree, and one tree is formed by constructing telecommunication lines from *Lincoln* to every other town.

13. We can represent the map of the fields by a graph $\mathcal{G}$ and represent breaking a hole in a wall by removing the corresponding edge from $\mathcal{G}$. To irrigate each field, there must be no cycles in the graph $\mathcal{H}$ that results from removing the edges corresponding to broken walls. Furthermore, in order to break as few walls as possible, the addition of an edge to $\mathcal{H}$ must create a cycle. Thus, $\mathcal{H}$ is a tree. Since $\mathcal{G}$ has 15 vertices and 26 edges, we want $\mathcal{H}$ to be a tree with 15 vertices and hence 14 edges. Thus, the farmer needs to break 12 walls.

15. There are many answers, one of which is given in the answer section of the textbook.

17. Consider a connected graph with n edges. By Exercise 65 of Section 4.2, the number m of vertices in the graph must satisfy $n \geq m - 1$. Hence $m \leq n + 1$. Also, in the connected graph below, there are exactly $n + 1$ vertices and n edges.

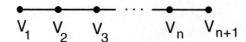

$$V_1 \quad V_2 \quad V_3 \quad \cdots \quad V_n \quad V_{n+1}$$

Thus the maximum possible number of vertices in a connected graph with n edges is $n+1$.

19. Since a cycle in T' would also be a cycle in T, T' has no cycles. Now suppose V_1 and V_2 are vertices in T'. Since V_1 and V_2 are also vertices in the tree T, there is a unique simple path between V_1 and V_2 using the edges in T. Since the vertex V has degree 1, any simple path containing V has V as one of its endpoints. Since the simple path between V_1 and V_2 does not have V as either endpoint, V and the edge e are not part of it. Hence, the simple path between V_1 and V_2 contains only edges and vertices from T'. Thus T' is connected and it is therefore a tree.

21. For $n > 2$, $\mathcal{K}_n$ has a cycle formed by the edges between any three vertices. On the other hand, $\mathcal{K}_1$ and $\mathcal{K}_2$ are both connected and have no cycles. Hence, only $\mathcal{K}_1$ and $\mathcal{K}_2$ are trees.

23. Look up in a chemistry reference book the chemical structure of butane and isobutane to obtain the tree structure of each as given in the answer section of the textbook.

25. A tree with 13 vertices has 12 edges, and so the sum of the degrees is 24. If there were 4 vertices of degree 3, 3 vertices of degree 4, and 6 vertices of degree 1, the sum of degrees would be $4 \cdot 3 + 3 \cdot 4 + 6 \cdot 1 = 30$.

27. See the answer section of the textbook for the three distinct trees with 3 labeled vertices.

29. Consider the first graph in the answer to Exercise 27 in the answer section of the textbook. Using the algorithm, we select the vertex of degree 1 that has the smallest label, which is the vertex with label 1. There is one edge on this vertex, and the other vertex on this edge has label 2. We put 2 in our list. Now we delete the vertex with label 1 and the edge on this vertex. This gives a tree with two vertices, and the algorithm is finished, giving a list with 2 as its only member. The algorithm is applied to the other two trees in the same manner, giving the lists in the answer section of the textbook.

31. Each iteration of step 2 of Prufer's algorithm gives the following.

Labels of Vertices of Degree 1	Label of the Vertex Deleted	List
1, 3, 4, 6, 7	1	2
3, 4, 6, 7	3	2, 2
2, 4, 6, 7	2	2, 2, 5
4, 6, 7	4	2, 2, 5, 5
6, 7	6	2, 2, 5, 5, 5

33. Following the process described in the exercise, we construct a tree as follows. First we have that L is the list 2, 2, 2, 2 and that N is $\{1, 2, 3, 4, 5, 6\}$. The smallest number in N that is not in L is 1, and so we construct an edge between 1 and the first number in L, which is 2. Then delete 2 from L and 1 from N. This gives the list L as 2, 2, 2 and N as $\{2, 3, 4, 5, 6\}$ and the following graph.

Again we choose the smallest number in N that is not in L, which is 3, and construct an edge between 3 and the first number in L, which is 2. We then delete 2 from L and 3 from N, which gives L as 2, 2 and N as $\{2, 4, 5, 6\}$. The graph is now as follows.

Continuing, we choose 4, the smallest number in N that is not in L, and construct an edge between 4 and 2, the first number in L. We then delete 2 from L and 4 from N, so that L is 2 and N is $\{2, 5, 6\}$. The graph is now as follows.

Continuing as before, we construct an edge between 5 and 2. Now L is exhausted and N is $\{2, 6\}$. The graph is as follows.

Joining the remaining two numbers in N yields the graph given in the answer section of the textbook.

35. Following the process described in Exercise 33, we have the following.

First Number in N Not in L	Added Edge on Labels	L	N
–	–	1, 2, 3, 2, 1	1, 2, 3, 4, 5, 6, 7
4	1, 4	2, 3, 2, 1	1, 2, 3, 5, 6, 7
5	5, 2	3, 2, 1	1, 2, 3, 6, 7
6	6, 3	2, 1	1, 2, 3, 7
3	3, 2	1	1, 2, 7
2	2, 1	exhausted	1, 7
	1, 7		

This process yields the tree given in the answer section of the textbook.

37. Following the process described in Exercise 33 and illustrated in Exercises 33 and 35 yields the graph given in the answer section of the textbook.

39. By definition a tree is connected, and by Theorem 5.3 the number of vertices is one more than the number of edges.

41. Suppose T is a graph that has no cycles and its number of vertices is one more than its number of edges. By Theorem 4.4, it is sufficient to show that there is a path between any two vertices or, equivalently, that T is connected. Suppose T is not connected. Consider the components of T. (See Exercise 57 of Section 4.2.) Since T is not connected, there is more than one. Each component is connected and has no cycles and is therefore a tree. By Theorem 5.3, the number of vertices in each component is one more than the number of edges. Thus the sum of the number of edges in the components is less than the number of edges in T, which is a contradiction.

43. Suppose T is a connected graph in which the removal of any edge results in a graph that is not connected. Suppose the connected graph T had a cycle. When an edge $\{U, V\}$ of that cycle is removed, the remaining portion of the cycle could still be used as an alternate path between U and V. Thus the resulting graph is still connected, which is a contradiction. So we conclude that T has no cycles. Now suppose that between two nonadjacent vertices V and U an edge e is added to the graph T, to form a new graph T_1. Since T is connected there is a simple path from V to U consisting of edges in T. This path along with e will form a cycle in T_1.

45. We shall proceed by mathematical induction on the number k of edges. Suppose $k = 1$. Since there are no loops in a tree, a tree with one edge has two vertices. Now assume the result holds for any tree with k or fewer edges, and let T be a tree with $k + 1$ edges. Delete an edge from T. The resulting graph has two components, T_1 and T_2, each of which is a tree with k_1 and k_2 edges, respectively. Applying mathematical induction to each of T_1 and T_2 gives that T_1 and T_2 have $k_1 + 1$ and $k_2 + 1$ vertices, respectively. Then the number of vertices in T is $(k_1 + 1) + (k_2 + 1) = k + 2$.

47. By Theorem 5.2 there is at least one vertex of degree 1 in a tree with at least two vertices. Hence, step 2 in Prufer's algorithm can be executed for a tree with at least two vertices. Furthermore, by Exercise 19 the deletion of the edge e and the vertex X from T as described in steps 2.2 and 2.3 of Prufer's algorithm results in a new tree. Thus, by Prufer's algorithm we can associate with each labeled tree a unique list of $n - 2$ integers from $\{1, 2, \ldots, n\}$. When each of the edges incident on V, except for the last one, is deleted, the label for V is placed in this list. Hence, if a vertex V in T has degree k, then the label for V appears $k - 1$ times in the list for T. If the last edge incident on V is deleted, then the label on the other vertex incident on the last edge is placed in the list. If the last edge incident on V is never deleted (which means V is one of the two remaining vertices in the tree when the algorithm finishes), the list is done. Hence, the list for T contains the label for V a total of $k - 1$ times. In particular, the labels for the vertices of degree 1 in T are precisely those that do not appear in the list for T.

Suppose L is the list for the tree T and $N = \{1, 2, \ldots, n\}$. When we choose the smallest number v in N that is not in L, we are choosing the vertex of degree 1 in T with the smallest label. Now suppose we delete both v from N and the first member of the list L. Then again choosing the smallest number in N that is not in L is the same as choosing the vertex of degree 1 with the smallest label in the tree resulting from the deletion from T of the vertex corresponding to v and the edge on it. Thus the ith vertex picked in the process described in Exercise 33 is the ith vertex picked in Prufer's algorithm. In the process described in Exercise 33, an edge is constructed where an edge is deleted in Prufer's algorithm. Hence the construction process in Exercise 33, when applied to the list for the tree T, reconstructs the tree T. In other words, Prufer's algorithm yields distinct lists for distinct trees. Note also that the construction process described in Exercise 33 generates from a list of length $n - 2$ a graph with n vertices and $n - 1$ edges. Since in this process, a label in N that is not in L corresponds to a vertex of degree 1, an edge drawn to a vertex of degree 1 cannot be part of a cycle. Thus the resulting graph is a tree. So Prufer's algorithm establishes a one-to-one correspondence between distinct trees on n labels and lists of $n - 2$ integers from $\{1, 2, \ldots, n\}$.

5.2 SPANNING TREES

1. Applying the breadth-first search algorithm as in Example 5.8 yields the following.

Vertices Included in $\mathcal{L}$	Label	Edges Included in $\mathcal{T}$
A	$0(-)$	none
B, C	$1(A)$	$\{A, B\}, \{A, C\}$
D	$2(B)$	$\{B, D\}$
E, F	$3(D)$	$\{D, E\}, \{D, F\}$
G	$4(E)$	$\{E, G\}$
I, H	$5(G)$	$\{G, H\}, \{G, I\}$
J	$6(H)$	$\{H, J\}$
K, L	$7(J)$	$\{J, K\}, \{J, L\}$
M	$8(K)$	$\{K, M\}$

This process yields the spanning tree given in the answer section of the textbook.

3. Applying the breadth-first search algorithm as in Example 5.8 and Exercise 1 yields the spanning tree given in the answer section of the textbook.

5. Applying the breadth-first search algorithm as in Example 5.8 and Exercise 1 yields the spanning tree given in the answer section of the textbook.

7. To satisfy the condition of being able to go from any building to any other and to avoid reinforcing more tunnels than necessary requires a minimal spanning tree. Applying the breadth-first search algorithm as in Example 5.8 and Exercise 1 yields the spanning tree given in the answer section of the textbook.

9. The answer is no, and there are many counterexamples, one of which is given in the answer section of the textbook.

11. Suppose e is an edge whose removal disconnects a connected graph $\mathcal{G}$ and that $\mathcal{T}$ is a spanning tree for $\mathcal{G}$. Removing e from $\mathcal{G}$ results in a graph $\mathcal{G}_1$ that is not connected. Thus there are two vertices V and U in $\mathcal{G}_1$ for which there is no path in $\mathcal{G}_1$ between them. Since $\mathcal{T}$ is a spanning tree for $\mathcal{G}$, there is a path between U and V using the edges of $\mathcal{T}$. Thus at least one edge of $\mathcal{T}$ cannot be in $\mathcal{G}_1$, and this must be the edge e. Thus e is an edge in $\mathcal{T}$.

13. There are many answers, and applying the breadth-first search algorithm as in Example 5.8 and Exercise 1 yields the spanning tree given in the answer section of the textbook.

15. We apply the breadth-first search algorithm as in Example 5.8 and Exercise 1. For convenience we will refer to a vertex by its number and use the term *bfs label* for the numerical label assigned by the breadth-first search algorithm. We begin with vertex

1 and assign it the bfs label 0 and cross out the number 1 in each adjacency list. Next we go to the adjacency list for vertex 1 and assign the vertices in it (2, 3, 5, 7, and 9) the bfs label 1. Then we cross out these numbers (2, 3, 5, 7, and 9) in each adjacency list. Then we go to the adjacency lists for the vertices 2, 3, 5, 7 and 9 and assign any uncrossed-out number the bfs label 2. The vertices assigned the bfs label 2 are 4, 6, and 8. Since the algorithm terminates with all the vertices labeled, the graph is connected.

17. As in Example 5.12, Prim's algorithm yields the following.

Edges Not Forming a Cycle	Respective Weights	Edge Included in T	Vertex Included in $\mathcal{L}$
—	—	none	A
a, b, d	$3, 4, 5$	a	B
b, c, d, e	$4, 3, 5, 6$	c	E
b, c, e, g, h	$4, 5, 6, 1, 3$	g	C
b, f, h	$4, 2, 3$	f	D

The algorithm terminates with a minimal spanning tree formed by the edges in T and the vertices in $\mathcal{L}$ having weight 9.

19. As in Example 5.12 and Exercise 17, Prim's algorithm yields the following.

Edges Not Forming a Cycle	Respective Weights	Edge Included in T	Vertex Included in $\mathcal{L}$
—	—	none	A
b, c, d	$6, 3, 4$	c	C
a, b, d, n	$4, 6, 4, 5$	a	B
d, n	$4, 5$	d	D
e	2	e	E
f, k, m	$2, 1, 4$	k	G
f, h, i, m	$2, 4, 2, 4$	f	H
g, i, m	$3, 2, 4$	i	I
j, m	$3, 4$	j	F

The algorithm terminates with a minimal spanning tree formed by the edges in T and the vertices in $\mathcal{L}$ having weight 21.

21. As in Example 5.12 and Exercises 17 and 19, a minimal spanning tree with weight 9 is formed as given in the answer section of the textbook.

23. As in Example 5.12 and Exercises 17 and 19, a minimal spanning tree with weight 21 is formed as given in the answer section of the textbook.

25. As in Example 5.13 and Exercises 17 and 19, a maximal spanning tree with weight 18 is formed as given in the answer section of the textbook.

27. As in Example 5.13 and Exercises 17 and 19, a maximal spanning tree with weight 31 is formed as given in the answer section of the textbook.

29. The situation can be modeled by a weighted graph where the bins are represented by vertices and the grain pipes are represented by weighted edges with the weight being the cost of building that pipe. Building the grain pipes at minimal cost is equivalent to finding a minimal spanning tree. As in Example 5.12 and Exercises 17 and 19, a minimal spanning tree can be formed as given in the answer section of the textbook.

31. For an edge to be part of every spanning tree, its removal from a graph must disconnect the graph, as shown in Exercise 11. There are many answers, one of which is given in the answer section of the textbook.

33. Prim's algorithm is modified so that in step 1 T is initialized to consist of the specified edge and $\mathcal{L}$ is to consist of the vertices of the specified edge. The rest of the algorithm continues in the same way. Then as in Example 5.12 and Exercises 17 and 19, a minimal spanning tree can be formed as in the answer section of the textbook.

35. Since it is possible to construct a connected weighted graph so that some edge is part of every minimal spanning tree (see Exercise 31), we can construct a connected weighted graph so that this edge is an edge of largest weight. An example is given in the answer section of the textbook.

37. If the package cost 26 cents to mail, the greedy algorithm would begin with choosing a stamp of largest value, namely the 22-cent stamp. This would leave 4 cents of postage needed and this would then be done by using four 1-cent stamps. However, the choice of two 13-cent stamps would use fewer stamps.

39. Kruskal's algorithm yields the following.

Minimum Weight of Edges Not in T	Edges of Minimum Weight Not Forming a Cycle	Edge Included in T
1	i, m	i
1	m	m
2	d, f, g	d
2	f, g	f
2	g	g
3	b, c, n	b
3	c, n	c
3	n	n
4	a	a

The algorithm terminates with a minimal spanning tree as given in the answer section of the textbook.

41. Kruskal's algorithm yields the following.

Minimum Weight of Edges Not in T	Edges of Minimum Weight Not Forming a Cycle	Edge Included in T
1	k	k
2	f, j	f
2	j	j
3	c, e, g	c
3	e, g	e
3	g	g
4	b, d, q	b
4	d, q	d
4	q	q
5	i, o, v	i
5	o	o

The algorithm terminates with a minimal spanning tree as given in the answer section of the textbook.

43. Modify step 1 of Kruskal's algorithm so that T is initialized to consist of the specified edge and S is initialized to consist of all the other edges of G. The rest of the algorithm remains the same. As in Exercises 39 and 41, the algorithm yields a minimal spanning tree as given in the answer section of the textbook.

45. Suppose G is a connected weighted graph in which all the weights are different. Suppose T_1 and T_2 are two different minimal spanning trees. Choose the edge e of smallest weight that is in one tree but not the other, say e is in T_2 but not in T_1. Adding the edge e to the tree T_1 creates a graph H with a cycle. Thus there is an edge f in T_1 in this cycle that is not in T_2. Since all the weights of edges are different, the weight of f is greater than the weight of e by our choice of e. Then, as in the proof of Prim's algorithm given in the textbook, the graph H_1 formed by deleting f from H is a spanning tree for G. Since the weight of f is greater than the weight of e, the weight of H_1 is less than the weight of T_1, which is a contradiction. Hence, there is only one minimal spanning tree.

5.3 DEPTH-FIRST SEARCH

1. Using the depth-first search algorithm as in Example 5.14, we have the following.

Vertex Included in $\mathcal{L}$	Depth-First Search Number	Predecessor	Unlabeled Adjacent Vertices
A	1	$-$	C, H
C	2	A	F, G
F	3	C	B, E, G, H
B	4	F	D, E
D	5	B	E
E	6	D	none, so backup to D then B, and then F to consider G and H
G	7	F	H
H	8	G	none

This now gives a depth-first numbering of the vertices for the graph.

3. We proceed as in Example 5.14 and Exercise 1 to obtain the following depth-first search numbering for the vertices. A has depth-first search number 1 and no predecessor, B has depth-first search number 2 and predecessor A, E has depth-first search number 3 and predecessor B, C has depth-first search number 4 and predecessor E, D has depth-first search number 5 and predecessor C, and H has depth-first search number 6 and predecessor D. Backing up is necessary at this point. Continuing, J has depth-first search number 7 and predecessor D, and I has depth-first search number 8 and predecessor J. Again backtracking is needed. Continuing, G has depth-first search number 9 and predecessor E, and F has depth-first search number 10 and predecessor G.

5. We proceed as in Example 5.14 and Exercise 1 to obtain the following depth-first search numbering for the vertices. A has depth-first search number 1 and no predecessor, C has depth-first search number 2 and predecessor A, E has depth-first search number 3 and predecessor C, B has depth-first search number 4 and predecessor E, and F has depth-first search number 5 and predecessor B. At this point it is necessary to backtrack. Continuing, J has depth-first search number 6 and predecessor C. Backtracking, D has depth-first search number 7 and predecessor A. Backtracking again, G has depth-first search number 8 and predecessor A, H has depth-first search number 9 and predecessor G, and I has depth-first search number 10 and predecessor H.

7. As in Example 5.14, the edges of the spanning tree are formed by using each vertex and its predecessor. Thus, from Exercise 1, we have the edges $\{A, C\}$, $\{C, F\}$, $\{F, B\}$,

$\{B, D\}$, $\{D, E\}$, $\{F, G\}$, and $\{G, H\}$. A diagram of the spanning tree is given in the answer section of the textbook.

9. As in Example 5.14, the edges of the spanning tree are formed by using each vertex and its predecessor. Thus from Exercise 3 we have the edges $\{A, B\}$, $\{B, E\}$, $\{E, C\}$, $\{C, D\}$, $\{D, H\}$, $\{D, J\}$, $\{J, I\}$, $\{E, G\}$, and $\{G, F\}$. A diagram of the spanning tree is given in the answer section of the textbook.

11. As in Example 5.14, the edges of the spanning tree are formed by using each vertex and its predecessor. Thus from Exercise 5 we have the edges $\{A, C\}$, $\{C, E\}$, $\{E, B\}$, $\{B, F\}$, $\{C, J\}$, $\{A, D\}$, $\{A, G\}$, $\{G, H\}$, and $\{H, I\}$. A diagram of the spanning tree is given in the answer section of the textbook.

13. For the depth-first search numbering used in Exercise 1, the back edges of the graph are those edges not used as part of the spanning tree in Exercise 7. Thus, the back edges are $\{A, H\}$, $\{F, E\}$, $\{B, E\}$, $\{G, C\}$, and $\{H, F\}$.

15. For the depth-first search numbering used in Exercise 3, the back edges of the graph are those edges not used as part of the spanning tree in Exercise 9. Thus, the back edges are $\{A, E\}$, $\{B, F\}$, $\{C, H\}$, and $\{C, I\}$.

17. For the depth-first search numbering used in Exercise 5, the back edges of the graph are those edges not used as part of the spanning tree in Exercise 11. Thus, the back edges are $\{A, I\}$ and $\{F, C\}$.

19. Because a bridge is part of every spanning tree (see Exercise 11 of Section 5.2), a bridge in a graph must be one of the edges in the spanning tree constructed by the depth-first search algorithm. Thus we begin by using the depth-first search algorithm to construct the spanning tree consisting of the edges $\{A, B\}$, $\{B, C\}$, and $\{C, D\}$. We delete in turn each of these edges in the spanning tree from the original graph to see if the resulting graph remains connected. (This can be done again by applying the depth-first search algorithm.) In this case we see that deleting in turn each of the edges $\{A, B\}$, $\{B, C\}$, and $\{C, D\}$ from the original graph results in a graph that is still connected. Thus there are no bridges in the graph.

21. We follow the same process as described in Exercise 19. The spanning tree constructed by the depth-first search algorithm will have edges $\{A, B\}$, $\{B, D\}$, $\{B, E\}$, $\{E, C\}$, and $\{C, F\}$. When the edge $\{B, E\}$ is deleted from the original graph, the depth-first search algorithm shows that the resulting graph is not connected, and so the edge $\{B, E\}$ is a bridge.

23. As in Example 5.15, the first step is to apply the depth-first search algorithm, and we will use that done in Exercise 1. Next we assign directions to the tree edges (see Exercise 7) by going from the lower-depth first search number to the higher and to the back edges (see Exercise 13) by going from the higher to the lower. This yields a directed graph, a diagram of which is given in the answer section of the textbook.

25. As in Example 5.15, the first step is to apply the depth-first search algorithm, and we will use that done in Exercise 4. Next we assign directions to the tree edges (see Exercise 10) by going from the lower depth-first search number to the higher and to the back edges (see Exercise 16) by going from the higher to the lower. This yields a directed graph, shown in the answer section of the textbook.

27. We proceed as in Example 5.15 and apply the depth-first search algorithm to the graph. This gives the following depth first-search numbering: A–1, B–2, D–3, C–4, E–5, G–6, F–7, H–8, L–9, J–10, K–11, I–12, M–13, O–14, and N–15. The tree edges $\{A, B\}$, $\{B, D\}$, $\{D, C\}$, $\{D, E\}$, $\{E, G\}$, $\{G, F\}$, $\{G, H\}$, $\{H, L\}$, $\{L, J\}$, $\{L, K\}$, $\{K, I\}$, $\{K, M\}$, $\{M, O\}$, and $\{O, N\}$ are assigned directions from the lower depth-first search number to the higher. The back edges (the remaining edges) are assigned directions from the higher depth-first search number to the lower. This yields a directed graph, which is given in the answer section of the textbook.

29. Suppose there is some vertex A which is not put in $\mathcal{L}$ (that is, no label is assigned to A). Since $\mathcal{G}$ is connected, there is a path between U and A. With U in $\mathcal{L}$ and A not in $\mathcal{L}$, we can find adjacent vertices V and W such that V is in $\mathcal{L}$ and W is not in $\mathcal{L}$. This is a contradiction, since in step 2 of the algorithm we must assign labels to all vertices adjacent to each vertex in $\mathcal{L}$ before the algorithm stops. Hence, every vertex of $\mathcal{G}$ is in $\mathcal{L}$, and the tree formed by the edges in $\mathcal{T}$ and vertices in $\mathcal{L}$ is a spanning tree for $\mathcal{G}$.

31. We begin the depth-first search at vertex 1 and assign it the depth-first search number 1. Since vertices 2 and 3 are adjacent to both vertex 1 and to each other, we can assign depth-first search number 2 to vertex 2 and 3 to vertex 3, or we could assign depth-first search number 2 to vertex 3 and 3 to vertex 2. Hence, there are two different depth-first search numberings of $\mathcal{K}_3$.

33. We begin the depth-first search at vertex 1 and assign it the depth-first search number 1. Since each of the remaining $n - 1$ vertices are adjacent to vertex 1, there are $n - 1$ choices for the vertex to receive depth-first search number 2. Then since each of the remaining $n - 2$ vertices are adjacent to the vertex with depth-first number 2, there are $n - 2$ choices for the vertex to have depth-first search number 3. Continuing in this fashion, we see that there are $(n - 1)!$ different depth-first search numberings for $\mathcal{K}_n$.

35. Apply depth-first search to a connected graph $\mathcal{G}$ to obtain a spanning tree $\mathcal{T}$. Suppose $\mathcal{C}$ is a cycle of $\mathcal{G}$. Then since $\mathcal{T}$ contains no cycles, some edge of $\mathcal{G}$ is in $\mathcal{C}$ but not in $\mathcal{T}$. This edge is then a back edge. Conversely, suppose e is a back edge. Then e is not an edge in $\mathcal{T}$, but the vertices on e are vertices in $\mathcal{T}$. By Theorem 5.5, adding e to $\mathcal{T}$ results in a graph with a cycle. Thus e is contained in a cycle of $\mathcal{G}$.

37. We begin by placing a queen in the $1, 1$ position. A placement of a queen anywhere else in column 1 is not permissible. We then move to column 2 and see that placing the queen in either the $2, 1$ or $2, 2$ positions is not permissible. Thus there is no solution to the two-queens problem.

39. We will use the notation of Example 5.16. We begin by placing a queen in the $1, 1$ position. Then going from top to bottom in column 2, we place a queen in position $3, 2$ because a queen in position $1, 2$ would result in two queens in the same row and a queen in position $2, 2$ would result in two queens in the same diagonal. Next we examine possible positions in column 3 from top to bottom. We see that position $1, 3$, position $2, 3$, position $3, 3$, and position $3, 4$ are not available because of the need to avoid having 2 queens in the same row or on the same diagonal. But a queen can be placed in position $5, 3$. Then we examine possible positions in column 4 from top to bottom. Position $1, 4$ is not available, but a queen can be placed in position $2, 4$. Finally examining the possible positions in column 5 from top to bottom indicates that we can place a queen in position $4, 5$. The placement of the queens is illustrated below.

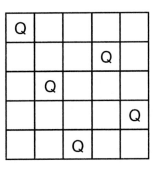

41. We will construct a sequence by adjoining the digits on the right and in numerical order. We begin with the sequence 1. We then adjoin the next digit, which according to our procedure will be 1. This gives the sequence 11, which does not satisfy the condition that nowhere in the sequence are there two adjacent subsequences that are identical. Then we backtrack to the sequence 1 and adjoin 2, giving the sequence 12. This sequence satisfies the desired condition, and so we adjoin 1 giving the sequence 121, which also satisfies the desired condition. So we adjoin 1 giving the sequence 1211, which does not satisfy the desired condition. So we backtrack to the sequence 121 and adjoin 2, giving the sequence 1212. But this does not satisfy the condition either, and so we backtrack to the sequence 121 and adjoin 3. The sequence 1213 does satisfy the condition, and so we adjoin 1 to the sequence giving 12131. This sequence also satisfies the condition, and so we adjoin 1, giving the sequence 121311, which does not. So we backtrack to the sequence 12131 and adjoin 2, giving 121312. This sequence satisfies the condition, and so we adjoin 1 giving 1213121, which also satisfies the condition. Because adjoining 1, 2, or 3 to this sequence does not satisfy the condition, we backtrack to the sequence 121312 and adjoin 2, but this sequence

99

1213122 is not acceptable. So we backtrack to 121312 and adjoin 3, giving 1213123, which satisfies the condition. Finally we adjoin 1, giving 12131231, which also satisfies the condition.

5.4 ROOTED TREES

1. The directed graph is a rooted tree since (i) if the directions of the edges are ignored, the resulting graph is a tree, and (ii) there is a unique vertex with indegree 0 and every other vertex has indegree 1.

3. The directed graph is not a rooted tree since there is more than one vertex with indegree 0.

5. The directed graph is not a rooted tree since, if the directions of the edges are ignored, the resulting undirected graph is not a tree.

7. The directed graph is not a rooted tree since, if the directions of the edges are ignored, the resulting undirected graph is not a tree.

9. As in Example 5.18, we have the graph given in the answer section of the textbook.

11. As in Example 5.18, we have the graph given in the answer section of the textbook.

13. The root is the object S-expression, and we draw two edges down from it with vertices to represent an atom and a list. We draw two edges down from an atom with vertices to represent a symbol and a number. We continue this pattern to produce the graph given in the answer section of the textbook.

15. The genealogical chart cannot be represented by a tree since a vertex representing a child born to Tom and Sue would have indegree greater than 1.

17. There are different ways to write this algorithm, one of which is given in the answer section of the textbook. Applying this algorithm to the graph in Exercise 17 gives the graph in the answer section of the textbook.

19. Since the root is to have indegree 0, the edges incident with the root must be assigned directions away from the root. Because the other vertices on these edges are to have indegree 1, all other edges on these vertices must be assigned directions away from these vertices. Continuing in this fashion, we see that there is only one way to assign directions to the edges of the tree so that it becomes a rooted tree with root at R.

21. (i) From the diagram, we see that the root is A.

(ii) Since the internal vertices are those with children, they are A, B, C, D, H, and I.

(iii) Since the terminal vertices are those without children, they are J, K, L, E, F, and G.

(iv) Since there is a directed edge from C to G, the parent of G is C.

(v) Since there are directed edges from B to D, E, and F, the children of B are D, E, and F.

(vi) Since D is on the simple directed path from the root to H, I, J, K, and L, the descendants of D are H, I, J, K, and L.

(vii) Since A, B, and D are on the simple directed path from the root to H, the ancestors of H are A, B, and D.

23. (i) From the diagram, we see that the root is E.

(ii) Since the internal vertices are those with children, the internal vertices are E, A, D, I, and J.

(iii) Since the terminal vertices are those without children, the internal vertices are B, K, G, F, H, and C.

(iv) Since there is a directed edge from D to G, the parent of G is D.

(v) Since there are no directed edges from B, there are no children of B.

(vi) Since D is on the simple directed path from the root to G, the descendant of D is G.

(vii) Since E and J are on the simple directed path from the root to H, the ancestors of H are E and J.

25. For a rooted tree to have as many internal vertices as possible, it needs to have as many vertices with children as possible, that is, as few terminal vertices as possible. An example is given in the answer section in the textbook.

27. The root will have three children representing the three possible outcomes of win, draw, or lose. Each of these three children will have three children representing again the three possible outcomes of win, draw, or lose. An example of such a rooted tree is given in the answer section of the textbook.

29. We follow the pattern in the second part of Example 5.21 in which there can be three outcomes to each weighing. Since we know the counterfeit coin is lighter, an example of such a decision tree is given in the answer section of the textbook.

31. Since the weight of the counterfeit coin is unknown, it is necessary to introduce an extra weighing to determine if the coin is heavier or lighter. One possible decision tree is given in the answer section of the textbook.

33. Since the length of a simple directed path from a vertex to itself is 0, the level of the root is 0.

35. Since the simple directed path from the root to F is A, B, F, the level of the vertex F is 2.

37. Since the simple directed path from the root C to H is C, H, the level of the vertex H is 1.

39. We first observe that if there is a tree edge from vertex A to vertex B in the depth-first search tree, then the depth-first search algorithm labels A before B. Hence, the depth-first search number of A is less than the one for B. Now suppose W is a descendant of V in the depth-first search tree. Then there exists a directed path of tree edges from V to W. Applying our observation, we see that the depth-first search numbers get larger for the vertices along the path from V to W. Hence, the depth-first search number of W is larger than the one for V.

5.5   BINARY TREES AND TRAVERSALS

1. As in Examples 5.24 through 5.27, we first form the binary tree with $a * b$ as the left child, $+$ as the root, and c as the right child. Then we replace the left child by the binary tree with a as the left child, $*$ as the root, and b as the right child. This produces the expression tree given in the answer section of the textbook.

3. As in Exercise 1, we first form the binary tree with $(a - b)/c$ as the left child, $*$ as the root, and $d + e/f$ as the right child. Then we replace the left child by the binary tree with $a - b$ as the left child, $/$ as the root, and c as the left child, and we replace the right child by the binary tree with d as the left child, $+$ as the root, and e/f as the right child. Continuing this process one more time yields the expression tree given in the answer section of the textbook.

5. As in Exercises 1 and 3, we form a binary tree with $a * (b * (c * (d * e + f) - g) + h)$ as the left child, $+$ as the root, and j as the right child. Next we replace the left child in the binary tree above by the binary tree with a as the left child, $*$ as the root, and $b * (c * (d * e + f) - g) + h$ as the right child. Then we replace the right child of this last binary tree by the binary tree with $b * (c * (d * e + f) - g)$ as the left child, $+$ as the root, and h as the right root. By continuing this process, we produce the expression tree given in the answer section of the textbook.

7. As in Example 5.23, we locate the left child B of A, and then we find all of the descendants of B. The left subtree of A is given in the answer section of the textbook.

9. As in Example 5.23 and Exercise 7, we locate the left child of C, which is E, and then we find all of the descendants of E. The left subtree of vertex C is given in the answer section of the textbook.

11. As in Example 5.23 and Exercise 7, we locate the left child of E, which is H, and then we find all of the descendants of H. The left subtree of vertex E is given in the answer section of the textbook.

13. As in Example 5.28, we visit the root (which is A), then we visit the left child (which is B), and then finally we visit the right child (which is C). This gives the preorder listing A, B, C of the vertices.

15. As in Example 5.28, we begin by visiting the root A, and then we visit its left child B. Then we visit the left child D of B. Since D has no left child, we visit the right child F of D. Since B has no right child, we visit the right child C of A. Then we visit the left child E of C. Since E has no left child, we visit the right child G of E. This gives the preorder listing A, B, D, F, C, E, G of the vertices.

17. As in Example 5.28 and Exercise 15, we have the preorder listing of the vertices given in the answer section of the textbook.

19. As in Example 5.31, we begin at the root and go to its left subtree, which consists of only the vertex B. After visiting B, we return to A and go to its right subtree, which consists only of the vertex C. After visiting C, we visit the vertex A. This gives the postorder listing B, C, A of the vertices.

21. As in Example 5.31 and Exercise 19, we begin at the root A and go to its left subtree, which has B as the root. Then we go to the left subtree of B, which has D as its root. Since D has no left subtree, we go to the right subtree of D, which has F as its root. Since F has no left subtree or right subtree, we visit F. Then we visit D. Since B has no right subtree, we visit B and go to the right subtree of A, which has C as its root. We then go to the left subtree of C, which has E as its root. Since E has no left subtree, we go to the right subtree of E, which has G as its root. Since G has no left or right subtree, we visit G. Then we visit E. Since C has no right subtree, we visit C. Finally we visit A. This gives the postorder listing F, D, B, G, E, C, A.

23. As in Example 5.31 and Exercises 19 and 21, we have the postorder listing of the vertices given in the answer section of the textbook.

25. As in Example 5.33, we begin at the root and go to the left subtree, which consists of only the vertex B. After visiting B, we visit A. Next we go to the right subtree of A, which consists only of the vertex C, which is visited. This gives the inorder listing B, A, C of the vertices.

27. As in Example 5.33 and Exercise 25, we begin at the root and go to its left subtree, which has the root B. Then we go to the left subtree of B, which has D as its root. Since D has no left subtree, we visit D and go to the right subtree of D, which has F as its root. Since F has no left subtree, we visit F, and since F has no right subtree, we next visit B. Since B has no right subtree, we visit A. Then we go to the right subtree of A, which has C as its root. Then we go the left subtree of C, which has E as its root. Since E has no left subtree, we visit E and go to the right subtree of E, which has G as its root. Since G has no left subtree, we visit G, and since G has no right subtree, we visit C. This gives the inorder listing of D, F, B, A, E, G, C of the vertices.

29. As in Example 5.33 and Exercises 25 and 27, we have the postorder listing of the vertices given in the answer section of the textbook.

31. We first construct the expression tree for the indicated expression. (See Exercise 1.) To find the Polish notation expression, we perform a preorder traversal on the expression tree. This results in the expression given in the answer section of the textbook.

33. We first construct the expression tree for the indicated expression. (See Exercise 3.) To find the Polish notation expression, we perform a preorder traversal on the expression tree. This results in the expression given in the answer section of the textbook.

35. We first construct the expression tree for the indicated expression. (See Exercise 5.) To find the Polish notation expression, we perform a preorder traversal on the expression tree. This results in the expression given in the answer section of the textbook.

37. We first construct the expression tree for the indicated expression. (See Exercise 1.) To find the reverse Polish notation expression, we perform a postorder traversal on the expression tree. This results in the expression given in the answer section of the textbook.

39. We first construct the expression tree for the indicated expression. (See Exercise 3.) To find the reverse Polish notation expression, we perform a postorder traversal on the expression tree. This results in the expression given in the answer section of the textbook.

41. We first construct the expression tree for the indicated expression. (See Exercise 5.) Then to find the reverse Polish notation expression, we perform a postorder traversal on the expression tree. This results in the expression given in the answer section of the textbook.

43. As in Example 5.30, we evaluate the Polish notation expression as follows.

$$+ \quad / \quad 4 \quad 2 \quad + \quad 5 \quad 6$$
$$+ \quad 2 \quad + \quad 5 \quad 6$$
$$+ \quad 2 \quad 11$$
$$13$$

45. As in Example 5.30, we evaluate the Polish notation expression as follows.

$$+ \quad * \quad 4 \quad / \quad 6 \quad 2 \quad - \quad + \quad 4 \quad 2 \quad 5$$
$$+ \quad * \quad 4 \quad 3 \quad - \quad + \quad 4 \quad 2 \quad 5$$
$$+ \quad 12 \quad - \quad + \quad 4 \quad 2 \quad 5$$
$$+ \quad 12 \quad - \quad 6 \quad 5$$
$$+ \quad 12 \quad 1$$
$$13$$

47. As in the textbook discussion in the section on postorder traversal, we evaluate the reverse Polish notation expression in the following manner.

$$
\begin{array}{ccccccccc}
4 & 5 & - & 7 & * & 2 & 3 & + & + \\
-1 & 7 & * & 2 & 3 & + & + & & \\
-7 & 2 & 3 & + & + & & & & \\
-7 & 5 & + & & & & & & \\
-2 & & & & & & & &
\end{array}
$$

49. As in the textbook discussion in the section on postorder traversal, we evaluate the reverse Polish notation expression in the following manner.

$$
\begin{array}{cccccccccc}
2 & 3 & + & 4 & 6 & - & - & 5 & * & 4 & + \\
5 & 4 & 6 & - & - & 5 & * & 4 & + & & \\
5 & -2 & - & 5 & * & 4 & + & & & & \\
7 & 5 & * & 4 & + & & & & & & \\
35 & 4 & + & & & & & & & & \\
39 & & & & & & & & & &
\end{array}
$$

51. Since $* + B\,D - A\,C$ is a Polish notation expression, it is the preorder listing of the vertices of an expression tree. Thus, $*$ is the root of an expression tree which has both a left subtree and a right subtree. Since the next item in the listing is an operation $+$, the operation $+$ is the root of the left subtree of the root $*$. Next we have that B is the left child for $+$, and D is the right child for $+$. Then we have that $-$ is the root of the right subtree for the root $*$. Finally we have that A is the left child for $-$, and C is the right child for $-$. This yields the expression tree given in the answer section of the textbook.

53. Since $A\,C * B - D +$ is a reverse Polish notation expression, it is the postorder listing of the vertices of an expression tree. Thus, $+$ is the root of an expression tree which has both a left subtree and a right subtree. Since an operand D immediately precedes $+$ in the postorder listing of the expression tree, the right subtree of $+$ is just the right child D. Then we see that the root of the left subtree of $+$ is $-$. Next we have that B is the right child of $-$, and the root of the left subtree of $-$ is $*$. Finally the left child of $*$ is A, and the right child of $*$ is C. This yields the expression tree given in the answer section of the textbook.

55. Since the preorder listing of the vertices is C, B, E, D, A, the first item C is the root. Because the inorder listing is B, E, C, A, D, the items before C in the inorder listing, B and E, are in the left subtree of C, and the items after C in the inorder listing, A and D, are in the right subtree of C. Thus B, E is the preorder listing of the vertices in the left subtree of C, which means that B is the root of the left subtree of C. Since B, E is the inorder listing of the left subtree of C, it follows that E is the right child of B. A similar argument for the right subtree of C yields the expression tree given in the answer section of the textbook.

105

57. Since the postorder listing of the vertices is E, B, F, C, A, D, the last item D is the root. Since the inorder listing of the vertices is E, B, D, F, A, C, the items appearing before D in the inorder listing, E and B, are in the left subtree of D, and the items appearing after D in the inorder listing, F, A, and C, are in the right subtree of D. Since E, B is the postorder listing of the vertices in the left subtree of D, B is the root of the left subtree of D. Then since E, B is the inorder listing of the vertices in the left subtree of D, we see that E is the left child of B. In a similar fashion, we see that since F, D, A is the postorder listing of the vertices in the right subtree of D, the root of this subtree is A. Since F, C, A is the inorder listing of the vertices of the right subtree of D, we have that F is the left child of A, and C is the right child of A. This yields the expression tree given in the answer section of the textbook.

59. To have the preorder listing be the same as the inorder listing, no vertex can have a left child. A binary tree with 7 vertices for which the preorder listing is the same as the inorder listing is given in the answer section of the textbook.

61. Since the root of a binary tree is listed first in the preorder listing and last in the postorder listing, a binary tree for which the preorder listing is the same as the postorder listing consists of a single vertex.

63. To have 1, 2, 3 as the postorder listing of the vertices of a binary tree means that 3 is the root. There are two possible ways to generate 1, 2 as the first two items in the postorder listing, and these are given in the answer section of the textbook.

65. Since the vertex X is a descendant of the vertex Y in a binary tree, there is a directed path from Y to X. Thus, X is in either the left subtree of Y or in the right subtree of Y. In the preorder traversal of a binary tree, the root of a subtree is listed before any vertex in the subtree. Hence, Y precedes X in the preorder listing of the vertices. In the postorder traversal of a binary tree the root of a subtree is listed after all the vertices in the subtree. Thus, X precedes Y in the postorder listing of the vertices.

67. The number of vertices in T_n is $2F_n - 1$, where F_n is the nth Fibonacci number. (See Section 2.5.) The proof is established by mathematical induction on n, where it is immediate that the result holds for $n = 1$ and $n = 2$. Now asssume that the result holds for $1, 2, \ldots, k$ with $k \geq 2$. We will show that it holds for $k + 1$. The tree T_{k+1} is composed of the left subtree T_k, the root, and the right subtree T_{k-1}. Hence, by the induction hypothesis, the number of vertices in T_{k+1} is $(2F_k - 1) + 1 + (2F_{k-1} - 1)$. Since $2F_k + 2F_{k-1} = 2(F_k + F_{k-1}) = 2F_{k+1}$, the result follows.

5.6 OPTIMAL BINARY TREES AND BINARY SEARCH TREES

1. Since the codeword 101 appears as the first part of the codeword 1011, this set of codewords does not have the prefix property.

3. Since the codeword 101 appears as the first part of the codeword 1010, this set of codewords does not have the prefix property.

5. Since 0 is a codeword, there can be no other codeword starting with 0. Thus, all other codewords must begin with 1. Since 11 is a codeword, 1 cannot be a codeword. Hence, all additional codewords must start with 1 and have length 2 or more. Since 11 and 10 are both codewords, it is not possible to have additional codewords. Thus, there is no set of 6 codewords with the prefix property that contains 0, 10, and 11.

7. Since 01 and a10 are both codewords, then by the prefix property we see that $a = 1$. Similarly, since 00, 01, and 101, and bc1 are codewords, we have that $b = 1$ and $c = 1$.

9. Since a terminal vertex is not on the simple directed path from the root to another terminal vertex, the set of codewords generated by the process described in Example 5.36 has the prefix property. Thus, we need a binary tree with two terminal vertices in which each vertex has 0 or 2 children. One possibility is given in the answer section of the textbook.

11. Since a terminal vertex is not on the simple directed path from the root to another terminal vertex, the set of codewords generated by the process described in Example 5.36 has the prefix property. Thus, we need a binary tree with four terminal vertices in which each vertex has 0 or 2 children. There are many possibilities and one is given in the answer section of the textbook.

13. Since a terminal vertex is not on the simple directed path from the root to another terminal vertex, the set of codewords generated by the process described in Example 5.36 has the prefix property. Thus, we need a binary tree with eight terminal vertices in which each vertex has 0 or 2 children. There are many possibilities, one of which is given in the answer section of the textbook.

15. We observe that the set of codewords has the prefix property. From the root, we construct the right child as a terminal vertex assigned the codeword 1. Next we construct the left child A of the root as an internal vertex. From A, we construct the left child as a terminal vertex assigned the codeword 00. Also from A, we construct the right child B as an internal vertex. From B, we construct the right child as a terminal vertex assigned the codeword 011 and the left child C as an internal vertex. Then from C we construct the left child and the right child as terminal vertices assigned the codewords 0100 and 0101, respectively. This binary tree is given in the answer section of the textbook.

17. We observe that the set of codewords has the prefix property. From the root we construct the left child as a terminal vertex assigned the codeword 0. Also from the root, we construct the right child A as an internal vertex. From A, we construct the left child as a terminal vertex assigned the codeword 10. Also from A, we construct the right child B as an internal vertex. From B, we construct the left child as a terminal vertex assigned the codeword 110 and the right child C as an internal vertex. From

C, we construct the left child and the right child as terminal vertices assigned the codewords 1110 and 1111, respectively. The binary tree is given in the answer section of the textbook.

19. Starting at the left end and looking for the individual strings 010, 111, 000, 110, and 10 in some sequence generates the codewords 111, 010, 10, 000, 010, and 110, which are decoded as BATMAN.

21. Starting at the left end and looking for the individual strings 111, 0, 1010, 1011, and 100 in some sequence generates the codewords 100, 1011, 1010, 100, and 1011, which are decoded as TONTO.

23. Because 0 is the first digit in the message string, we start at the root and go to the left child. Since this left child is a terminal vertex with the assigned codeword G, we have G as the first decoded character of the message. We now return to the root and start with the second digit of the message string. Since this second digit is a 1, we go to the right child. Since this right child is an internal vertex, we use the next digit in the message string, which is 1. Being a 1 means that from there we go the right child, which is again an internal vertex. Thus, we use the next digit again, which is a 1, and so we go the next right child. Since this right child is a terminal vertex with the assigned codeword O, we have O as the second decoded character of the message. Now we return to the root to begin with the next unused digit in the message string which is 0. Continuing in this fashion, we find the decoded message to be GOGO.

25. Since the first digit of the message string is a 0, we start at the root and go to its left child. Since this vertex is not a terminal vertex, we go to its right child because the next digit of the message string is a 1. Again this vertex is not a terminal vertex, and so we go to its right child because the next digit of the message is a 1. Since this vertex is a terminal vertex with the assigned codeword T, the first decoded character of the message is T. We return to the root and begin the process again using the fourth digit in the message string. As before, we follow the directed path generated by the 0s and 1s until we reach a terminal vertex. In this case, the three digits of 1 direct us to go right three times, reaching a terminal vertex with codeword H. Thus, H is the second decoded character of the message. Continuing in this manner, we find the decoded message to be THEHATS.

27. Since characters are represented by 8 digit strings in ASCII code, we decompose the message string into individual strings of length 8: 01000100, 01001111, and 01000111. Using an ASCII table, we decode these as DOG.

29. Since characters are represented by 8 digit strings in ASCII code, we decompose the message string into individual strings of length 8: 01010001, 01010101, 01001001, 01000101, and 01010100. Using an ASCII table, we decode these as QUIET.

31. As in Examples 5.38 and 5.39, each iteration of step 2 of Huffman's optimal binary tree algorithm yields the following.

Label of Root of New Binary Tree	Label of Root of Left Child	Label of Root of Right Child	Labels of Roots of New Trees in $\mathcal{S}$
6	2	4	6, 6, 8, 10
12	6	6	8, 10, 12
18	8	10	12, 18
30	12	18	30

This yields the optimal binary tree given in the answer section of the textbook.

33. As in Examples 5.38 and 5.39, each iteration of step 2 of Huffman's optimal binary tree algorithm yields the following.

Label of Root of New Binary Tree	Label of Root of Left Child	Label of Root of Right Child	Labels of Roots of New Trees in $\mathcal{S}$
5	1	4	5, 9, 16, 25, 36
14	5	9	14, 16, 25, 36
30	14	16	25, 30, 36
55	25	30	36, 55
91	36	55	91

This yields the optimal binary tree given in the answer section of the textbook.

35. As in Example 5.41, the smallest maximum number of comparisons needed to merge together the lists is shown below.

Number of Items in Two Lists to be Merged	Maximum Number of Comparisons	Number of Items in Remaining Lists
20, 30	49	50, 40, 50
40, 50	89	90, 50
50, 90	139	14

Thus the optimal merge pattern uses at most 277 comparisons.

37. As in Example 5.41, the smallest maximum number of comparisons needed to merge together the lists is shown below.

Number of Items in Two Lists to be Merged	Maximum Number of Comparisons	Number of Items in Remaining Lists
20, 40	59	60, 60, 70, 80, 120
60, 60	119	120, 70, 80, 120
70, 80	149	150, 120, 120
120, 120	239	240, 150
240, 150	389	390

Thus the optimal merge pattern uses at most 955 comparisons.

39. We first construct an optimal binary tree with weights 40, 35, 20, and 5. As in Examples 5.38 and 5.39, each iteration of step 2 of Huffman's optimal binary tree algorithm yields the following.

Label of Root of New Binary Tree	Label of Root of Left Child	Label of Root of Right Child	Labels of Roots of New Trees in $\mathcal{S}$
25	5	20	25, 35, 40
60	25	35	40, 60
10	40	60	100

Thus the following optimal binary tree is constructed.

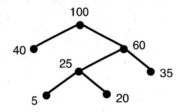

From this we can construct the following coding tree.

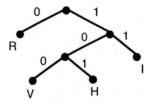

Hence the assignment of codewords given in the answer section of the textbook minimizes the number of bits needed to send a message.

41. We proceed as in Exercise 39. First, we construct an optimal binary tree with weights 125, 100, 75, 40, 60, 180, 20, 120, 150, and 130. This binary tree is transformed into a coding tree with 0 assigned when going left from a root and 1 assigned when going right from a root. In this way, each terminal vertex is assigned a codeword to be used for the numbers 1, 2, 3, 4, 5, 6, 7, 8, 9, 10, respectively. This assignment is given in the answer section of the textbook.

43. The proof of the theorem is by induction on the number n of terminal vertices. When $n = 1$, the binary tree consisting of only a root is a binary tree with 1 vertex in which every vertex has 0 or 2 children. Now suppose the theorem is true for k, and we will prove it for $k + 1$. Thus we assume that there is a binary tree T with k terminal vertices in which each vertex has 0 or 2 children. Suppose V is a terminal vertex. Then construct a binary tree T_1 from T by attaching two new vertices (one a left child, and one a right child). Then T_1 is a binary tree with $k + 1$ terminal vertices, and each vertex has 0 or 2 children.

45. Suppose the distance d_i from the root to w_i is less than the distance d_j from the root to w_j. Let T_1 be the binary tree obtained from T by interchanging the weights w_i and w_j. Thus, the terminal vertices w_i and w_j in the binary tree T_1 contribute the terms $d_j w_i$ and $d_i w_j$ to the weight of T_1, and all the other terminal vertices of T_1 contribute the same term as they do for the weight of T. Hence, the weight of T_1 subtracted from the weight of T is

$$d_i w_i + d_j w_j - d_j w_i - d_i w_j = (d_j - d_i)(w_j - w_i),$$

which is positive. Thus, T_1 is a binary tree of smaller weight than T, which is an optimal binary tree, and this is a contradiction.

47. Suppose T is an optimal binary tree for the weights $w_1 + w_2, w_3, \ldots, w_k$ and that T_1 is the binary tree obtained from T by replacing the terminal vertex $w_1 + w_2$ by a binary tree with two children w_1 and w_2. Since there are only a finite number of binary trees with k terminal vertices and only a finite number of ways to assign the k weights to these k terminal vertices, there is an optimal binary tree T_2 for the weights $w_1, w_2, \ldots, w_k$. By Exercise 46, we may assume that w_1 and w_2 are children of the same parent V in T_2. Removing the terminal vertices w_1 and w_2 from the tree T_2 results in a new binary tree T_3 with V as a terminal vertex. We now assign the weight $w_1 + w_2$ to the terminal vertex V in T_3. Let d be the distance from the root to V in T_2, and hence also in T_3. Thus, the terminal vertices w_1 and w_2 contribute $(d+1)w_1 + (d+1)w_2$ to the weight of T_2, and vertex V contributes $d(w_1 + w_2)$ to the weight of T_3. Hence, the weight of T_2 is the sum of the weight of T_3 and $w_1 + w_2$. In a similar fashion, the weight of T_1 is the sum of the weight of T and $w_1 + w_2$. Since T is an optimal binary tree for $w_1 + w_2, w_3, \ldots, w_k$, we have that the weight of T is less than or equal to the weight of T_3. If the weight of T is less than the weight of T_3, then the weight of T_1 is less than the weight of T_2, which is a contradiction since T_2 is an optimal binary tree for $w_1, w_2, \ldots, w_k$. Thus the weight of T is the weight of T_3, and so the weight of T_1 is the weight of T_2. Hence, T_1 is an optimal binary tree for the weights $w_1, w_2, \ldots, w_k$.

49. As in Example 5.44, we begin by assigning the root the first word in the given list, which is list. Since the second word in the given list, static, is greater than list, we go right. Since list has no right child, we construct a right child for list and assign static to this right child. The third word in the given list is or, which is greater than list.

111

Thus, we go to the right child static of the root. Since or is less than static, we go left from static. But since static has no left child, we construct a left child for static and assign or to this left child. The fourth word in the given list is begin, which is less than the root list, and so we go left. Since list has no left child, we construct a left child for list and assign begin to this left child. In a similar manner, the fifth word this in the given list is added to the tree as a right child of static. Likewise, we see that endl is added to the tree as a right child of begin. Continuing in this manner, we obtain the binary search tree given in the answer section of the textbook.

51. As in Example 5.44 and Exercise 49, we begin by assigning the root the first number in the list, which is 14. Since the next number in the list, 17, is greater than 14, we go right from the root. Since the root 14 has no right child, we construct a right child for 14 and assign 17 to this right child. The third number in the list is 3, which is less than the root 14, and so we go left from the root. Since the root 14 has no left child, we construct a left child for 14 and assign 3 to this left child. The fourth number in the list is 6, which is less than the root 14. Hence, we go to the left child 3 of the root 14. Since 6 is greater than 3, we go right from 3. But since 3 has no right child, we construct a right child of 3 and assign 6 to this right child. In a similar fashion, 15 is added to the tree as a left child of 17, and 1 is added to the tree as a left child of 3. Continuing in this manner, we obtain the binary search tree in the answer section of the textbook.

53. By using the decimal values of the symbols and proceeding as in Example 5.44 and Exercises 49 and 51, we construct the binary search tree given in the answer section of the textbook.

55. As in Example 5.46, we begin by comparing filenames to the root list of the binary search tree in Exercise 49. Since filenames is smaller, we go to the left child begin of the root and make another comparison. Since filenames is larger, we go to the right child endl of begin and make still another comparison. Since filenames is again larger, we go to the right child heap of endl. Comparing filenames and heap, we go to the left child of heap. Since there is none, filenames is not in the binary search tree. Furthermore, filenames would be added to the binary search tree as a left child of heap. This is illustrated by the diagram given in the answer section of the textbook.

57. As in Example 5.46 and Exercise 55, we begin by comparing 4 to the root 14 of the binary search tree of Exercise 51. Since 4 is smaller, we go to the left child 3 of the root and make another comparison. Since 4 is larger, we go to the right child 6 of 3 and make still another comparison. Since 4 is smaller than 6, we go to the left child 5 of 6. Comparing 4 and 5, we go to the left child of 5. Since there is none, 4 is not in the binary search tree. Furthermore, 4 would be added to the binary search tree as a left child of 5. This is illustrated by the diagram given in the answer section of the textbook.

59. By using the decimal values of the symbols and proceeding as in Example 5.46 and Exercises 55 and 57, we find that $>$ is not in the binary search tree. Furthermore, $>$

would be added to the binary search tree as a left child of ?. This is illustrated by the diagram in the answer section of the textbook.

61. A binary search tree with this kind of assignment need not be a binary search tree for the list as illustrated by the tree shown below.

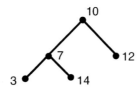

63. Deletion of the terminal vertex parameter from the binary search tree of Exercise 49 is accomplished by deleting the vertex parameter and the edge on vertices or and parameter. This results in the binary search tree given in the answer section of the textbook.

65. Deletion of root 22 from the binary search tree is accomplished by deleting the vertex 22 and the edge on the vertices 14 and 22. This results in the binary search tree given in the answer section of the textbook.

67. Deletion of the vertex 3 from the binary search tree of Exercise 48 is accomplished by deleting the vertex 3 and the edge on the vertices 3 and 6, and then by replacing the vertex 3 by the right subtree of 3, which is just the vertex 6. This results in the binary search tree given in the answer section of the textbook.

69. Deletion of the vertex 3 from the binary search tree of Exercise 51 is accomplished as follows. The largest item in the left subtree of 3 is 2. Since 2 has no left child, delete the vertex 2 and the edge on the vertices 2 and 1, and then replace 3 by 2. This results in the binary search tree given in the answer section of the textbook.

71. Deletion of the vertex variable from the binary search tree of Exercise 50 is accomplished as follows. The largest item in the left subtree of variable is template. Since template has no left child, delete the vertex template and the edge on the vertices template and string, and then replace variable by template. This results in the binary search tree given in the answer section of the textbook.

73. Deletion of the vertex 4 from the binary search tree of Exercise 52 is accomplished as follows. The largest item in the left subtree of 4 is 2. Since 2 has no left child, delete the vertex 2 and the edge on the vertices 2 and 4, and then replace 4 by 2. This results in the binary search tree given in the answer section of the textbook.

SUPPLEMENTARY EXERCISES

1. As in Example 5.5, this situation can be represented by a graph in which the vertices correspond to informants and an edge joins two vertices when the corresponding informants know about the same meeting place. This graph will be a tree with 6 edges and hence 7 vertices. Thus there will need to be 6 meeting places.

3. When $n = 1$, $\mathcal{K}_{m,n}$ is a connected graph with no cycles. However, when $n > 1$, $\mathcal{K}_{m,n}$ is connected but contains cycles.

5. This is immediate when $k = 0$, and so we assume $k > 0$. Let U be a vertex with degree k, and let $V_1, V_2, \ldots, V_k$ be the vertices adjacent to U. For each i, build a simple path with a maximal number of edges starting at U, then using V_i, and ultimately ending at W_i. Since the simple path has a maximal number of edges, W_i has degree 1. Furthermore, since there are no cycles in a tree, $W_1, W_2, \ldots, W_k$ are all distinct from each other.

7. Suppose $d_1 + d_2 + \cdots + d_n = 2n - 2$, where each d_i is a positive integer. If each $d_i > 1$, then $d_1 + d_2 + \cdots + d_n \geq 2n$, which is greater than $2n - 2$. Thus some $d_i = 1$. If each $d_i = 1$, then $d_1 + d_2 + \cdots + d_n = n$, which is less than $2n - 2$ when $n \geq 3$. Hence, when $n \geq 3$, some $d_i > 1$.

9. A spanning tree for the graph in the exercise contains 3 edges, is connected, and contains no cycles. The possibilities are shown in the answer section of the textbook.

11. Since $\mathcal{G}$ is a connected graph with 10 vertices, each spanning tree of $\mathcal{G}$ has 10 vertices and 9 edges. Thus 10 edges can be removed and still leave a connected graph. By Exercise 17 of Section 5.1, we know that a connected graph with 8 edges can have at most 9 vertices. Thus 10 is the maximum number of edges that can be removed from $\mathcal{G}$ so that the resulting graph remains connected.

13. For the connected weighted graph below, the spanning tree with edges of weights 1, 3, and 5 has the same weight as the spanning tree with edges of weights 2, 3, and 4.

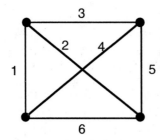

15. We can use Prim's algorithm as modified in Exercise 32 of Section 5.2 to find a minimal spanning tree containing e.

17. Kruskal's algorithm is modified so that, in step 1, $\mathcal{T}$ is initialized to consist of the two specified edges and $\mathcal{S}$ is initialized to consist of all the edges of $\mathcal{G}$ except for the two specified edges. The rest of the algorithm continues in the same way.

19. We proceed as in Exercise 7 of Section 5.3 using the depth-first search numbering of Exercise 18. The spanning tree for the first graph is given in the answer section of the textbook. The spanning tree for the second graph is as follows.

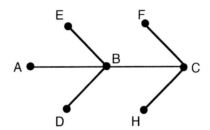

21. Suppose e is an edge on vertices A and B in a connected graph $\mathcal{G}$. Assume that e is a bridge. Suppose there is a path $\mathcal{P}$ in $\mathcal{G}$ from A to B which does not contain e. Then any path between two vertices U and V that contains e can be modified by replacing e with $\mathcal{P}$ so that it remains a path between U and V. Thus removing e from $\mathcal{G}$ results in a graph that is still connected, which contradicts that e is a bridge. Now assume that every path from A to B includes e. If e is removed from $\mathcal{G}$, then in the resulting graph there is no path from A to B, and the graph resulting from the deletion of e is not connected. So e is a bridge.

23. We will use the notation of Example 5.16. We begin by placing a queen in the $1, 1$ position. Then going from top to bottom in column 2, we place a queen in position $3, 2$ because a queen in position $1, 2$ would result in two queens in the same row and a queen in position $2, 2$ would result in two queens in the same diagonal. Next we examine possible positions in column 3 from top to bottom. We see that position $1, 3$, position $2, 3$, position $3, 3$, and position $3, 4$ are not available because of the need to avoid having 2 queens in the same row or diagonal. But a queen can be placed in position $5, 3$. Then we examine possible positions in column 4 from top to bottom. Position $1, 4$ is not available, but a queen can be placed in position $2, 4$. Looking from top to bottom for possible positions in column 5, we see that a queen can be placed in position $4, 5$. Finally we place a queen in position $6, 6$. A diagram showing the placement of the queens is given in the answer section of the textbook.

25. Extending the idea used to do Exercise 25 in Section 5.4, we construct the rooted tree given in (a) which has $n - 1$ internal vertices and 1 terminal vertex. We then construct

115

from this rooted tree a rooted tree which has $n - 2$ internal vertices and 2 terminal vertices by deleting the root and the edge on the root and by adding a child and edge at S. This is given in (b). We again delete the root and the edge on it and add a child and edge at S which results in the binary tree given in (c). We continue this recursive process until we arrive at a rooted tree with one internal vertex and $n - 1$ terminal vertices as given in (d).

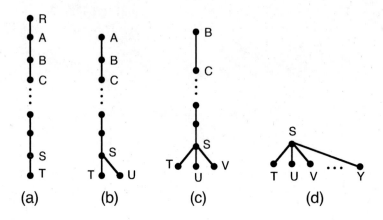

27. In a rooted binary tree, every vertex except the root is a child. Consider a rooted binary tree in which there are p internal vertices, each with exactly q children. Since each internal vertex has exactly q children and there are p internal vertices, there are pq vertices that are children. By including the root, we have $pq + 1$ vertices in all.

29. We proceed by mathematical induction on n. Suppose $n = 1$. Then there are 1, 2, or 3 coins. When there is 1 coin, then that coin is the light one. When there are 2 coins, then using the balance scale to weigh the 2 coins will indicate the light coin. When there are 3 coins, we use the balance scale to weigh two coins (one on each side). If the two sides balance, then the third coin is the light one. Otherwise, as in the case of 2 coins, the balance scale will indicate the light coin. Thus the light coin can be found with at most one weighing.

We now suppose that at most n weighings are needed for 3^n or fewer coins, and we will show that at most $n + 1$ weighings are needed for 3^{n+1} or fewer coins. If there are 3^n coins or fewer, the induction hypothesis implies that at most n (and, hence, at most $n + 1$) weighings are needed. Suppose there are more than 3^n coins but not more than $2 \cdot 3^n$ coins. If the number of coins is even, divide the coins into two sets A and B containing the same number of coins. If the number of coins is odd, divide the coins into two sets A and B containing the same number of coins and a third set C containing a single coin. In either case, weigh A on the right side of the scale and B on the left. If the balance tips down on the left, then A is a set of at most 3^n coins that contains the light coin. Applying the induction hypothesis to A, we can find the light coin in at most n weighings. So overall, at most $n+1$ weighings are needed to find the light coin. A similar argument works with B if the balance tips down on the right, and if A and B balance, then the single coin in C must be the light one.

Finally, suppose that there are more than $2 \cdot 3^n$ coins. In this case, we can find two sets of 3^n coins A and B and a set C containing fewer than 3^n coins. Again we weigh A on the right side and B on the left side. If the balance tips down on one side, the argument is the same as in the previous case. Otherwise, C is a set of fewer than 3^n coins that contains the light coin, and so by applying the induction hypothesis to C, the light coin can be found in at most $n+1$ weighings. This establishes the result for at most 3^{n+1} coins, and so completes the induction argument.

31. Suppose $\mathcal{H} = \{D_1, D_2, \ldots, D_m\}$ is the set of all m descendants of a vertex V relative to the depth-first search tree $\mathcal{T}$. Then for each $i = 1, 2, \ldots, m$, there is a directed path $\mathcal{P}_i$ from V to D_i using the (directed) edges of $\mathcal{T}$. Since each vertex other than V on each $\mathcal{P}_i$ is a descendant of V, the collection of vertices other than V on these paths is precisely the set $\mathcal{H}$. Furthermore, the vertices in $\mathcal{H}$ all have different depth-first search numbers, and the algorithm assigns each one a number larger than k. Since there are no directed paths from V using the edges of $\mathcal{T}$ other than those in $\mathcal{H}$, after the algorithm assigns V the depth-first search number k, the vertices in $\mathcal{H}$ must be assigned depth-first search numbers before any other vertices are assigned numbers. Hence the vertices in $\mathcal{H}$ are assigned the depth-first search numbers $k+1, k+2, \ldots, k+m$.

33. Suppose e is an edge incident on the vertices U and V. If e is a tree edge directed from U to V, then U is an ancestor of V. Otherwise, e is a back edge, and by Exercise 32 one of the vertices on e is an ancestor of the other.

35. Suppose for some child C of A there is no back edge between C or any of its descendants and an ancestor of A. Since A is not the root, there is a vertex U which is the parent of A. Suppose there is a simple path $\mathcal{P}$ between U and C which does not include A, and which is of maximum length with respect to these properties. In the path $\mathcal{P}$, the edge on C is either a tree edge or a back edge. If it is a tree edge, then the next vertex in $\mathcal{P}$ after C is a descendant C_1 of C. The next edge on C_1 in the path $\mathcal{P}$ is again either a tree edge or a back edge. If it is also a tree edge, then the next vertex in $\mathcal{P}$ after C_1 is a descendant C_2 of C. Since U is not a descendant of C, we must eventually come to a vertex D (which is C or one of its descendants) where the next edge e on D in the path $\mathcal{P}$ is a back edge. By Exercise 32, the other vertex V on e is an ancestor of D, since if D were the ancestor of V, the maximality of the length of $\mathcal{P}$ would be contradicted. By hypothesis V is not an ancestor of A. Since A is not on the path $\mathcal{P}$ and there is a unique path from D back through parents to the root, V must be one of the descendants of A. This contradicts the assumption that $\mathcal{P}$ is simple. Hence, every path between U and C must contain A. By Exercise 36 of Section 5.3, vertex A is an articulation point for $\mathcal{G}$.

Conversely, suppose that A is an articulation point for $\mathcal{G}$. Then there are distinct vertices U and V other than A such that every path between U and V contains A. If neither of U and V is a descendant of A, then the two paths consisting of the tree edges from the root to each of U and V do not contain A. These two paths can be joined together to form a path between U and V not containing A, which is a contradiction. Hence, at least one of U and V, say U, is a descendant of A. Then there is a vertex C that is a child of A such that either $U = C$ or U is a descendant of C. Suppose now that there is a back edge between C or one of its descendants, say X, and an ancestor of A, say Y. Since both X and U are descendants of C or equal to C, there is a path $\mathcal{P}_1$ between X and U not containing A. First suppose V is not a descendant of A. Then since Y is also not a descendant of A, we can find as before a path $\mathcal{P}_2$ between V and Y not containing A. Then by combining the path $\mathcal{P}_2$ with the back edge on Y and X and then with the path $\mathcal{P}_1$, we have a path between U and V that does not contain A, which is a contradiction. Thus, V is also a descendant of A. If V were a descendant of C, then there would be a path between U and V not containing A, which is a contradiction. Hence, V is not a descendant of C. Thus there is a vertex D distinct from C such that D is a child of A and $V = D$ or V is a descendant of D. We can suppose there is a back edge between D or one of its descendants, say W, and an ancestor of A, say Z, since otherwise the condition we want is satisfied (for D instead of C). As before, there is a path $\mathcal{P}_3$ between Y and Z that does not contain A and a path $\mathcal{P}_4$ between W and V that does not contain A. Thus, the path formed by combining the path $\mathcal{P}_1$, the back edge on X and Y, the path $\mathcal{P}_3$, the back edge on Z and W, and the path $\mathcal{P}_4$ is a path between U and V that does not contain A, which is a contradiction. Hence, there is no back edge between C or any of its descendants and an ancestor of A.

37. From Exercise 33 of Section 5.4, the level of a vertex is the length of the simple directed path from the root to that vertex. We proceed by mathematical induction on h. When $h = 1$, the terminal vertices are the children of the root, and there will be at most two of them. Assume the result is true for h and that $\mathcal{T}$ is a binary tree of height $h + 1$. By deleting the terminal vertices of $\mathcal{T}$, we have a binary tree $\mathcal{T}_1$ of height h. By induction $\mathcal{T}_1$ has at most 2^h terminal vertices. Since each of these has at most 2 children in $\mathcal{T}$, there will be at most $2^h \cdot 2$ terminal vertices in $\mathcal{T}$.

39. One possible answer is given in the answer section of the textbook.

41. The associative law for multiplication is $a*(b*c) = (a*b)*c$. The respective expression trees for $a * (b * c)$ and $(a * b) * c$ are as follows.

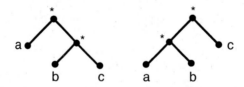

For Polish notation, we perform the preorder traversal on each expression tree and obtain $* a * b c$ and $* * a b c$, respectively. For reverse Polish notation, we perform the postorder traversal on each expression tree and obtain $a b c * *$ and $a b * c *$, respectively.

43. Because none of the codewords 00, 01, 100, 1010, 1011, and 11 is the first part of any other codeword, the given set of codewords has the prefix property.

45. We proceed as we did in Exercise 15 of Section 5.6. This yields the binary tree given in the answer section in the textbook.

47. We proceed as we did in Exercise 49 of Section 5.6. This yields the binary search tree given in the answer section in the textbook.

Chapter 6

Matching

6.1 SYSTEMS OF DISTINCT REPRESENTATIVES

1. There are two, namely 1, 2, 3 and 2, 3, 1.

3. There are $3! = 6$, the number of permutations of $\{1, 2, 3\}$.

5. There are none, since the union of the 1st, 2nd, 4th, and 5th sets has only 3 elements.

7.

I	union
$\varnothing$	$\varnothing$
$\{1\}$	$\{1, 2, 4\}$
$\{2\}$	$\{2, 4\}$
$\{3\}$	$\{2, 3\}$
$\{4\}$	$\{1, 2, 3\}$
$\{1, 2\}$	$\{1, 2, 4\}$
$\{1, 3\}$	$\{1, 2, 3, 4\}$
$\{1, 4\}$	$\{1, 2, 3, 4\}$
$\{2, 3\}$	$\{2, 3, 4\}$
$\{2, 4\}$	$\{1, 2, 3, 4\}$
$\{3, 4\}$	$\{1, 2, 3\}$
$\{1, 2, 3\}$	$\{1, 2, 3, 4\}$
$\{1, 2, 4\}$	$\{1, 2, 3, 4\}$
$\{1, 3, 4\}$	$\{1, 2, 3, 4\}$
$\{2, 3, 4\}$	$\{1, 2, 3, 4\}$
$\{1, 2, 3, 4\}$	$\{1, 2, 3, 4\}$

Because the number of elements in each set in the second column is greater than or equal to the number of elements in the corresponding set in the first column, a system of distinct representatives exists.

9.

I	union
$\varnothing$	$\varnothing$
$\{1\}$	$\{1\}$
$\{2\}$	$\{1,2\}$
$\{3\}$	$\{1,2,3\}$
$\{4\}$	$\{1,3\}$
$\{1,2\}$	$\{1,2\}$
$\{1,3\}$	$\{1,2,3\}$
$\{1,4\}$	$\{1,3\}$
$\{2,3\}$	$\{1,2,3\}$
$\{2,4\}$	$\{1,2,3\}$
$\{3,4\}$	$\{1,2,3\}$
$\{1,2,3\}$	$\{1,2,3\}$
$\{1,2,4\}$	$\{1,2,3\}$
$\{1,3,4\}$	$\{1,2,3\}$
$\{2,3,4\}$	$\{1,2,3\}$
$\{1,2,3,4\}$	$\{1,2,3\}$

Because $I = \{1,2,3,4\}$ contains 4 elements but $S_1 \cup S_2 \cup S_3 \cup S_4$ contains only 3 elements, no system of distinct representatives exists.

11. $I = \{3\}$, union $= \varnothing$

13. $I = 1,2,3,4,5$, union$= \{1,2,3,4\}$

15. $I = \{1,3,4,6\}$, union$= \{2,5,7\}$

17. There are $n!$, the number of permutations of $\{1,2,...,n\}$.

19. There are none, since $S_1 \cup S_2 \cup \cdots \cup S_n$ has fewer than n elements.

21. Amy, Burt, Dan, and Edsel like only the 3 flavors chocolate, banana, and vanilla.

23. Let $x_1, x_2, \ldots, x_n$ be a system of distinct representatives. There exists an element y in the union of the sets S_i not equal to any element x_i. Suppose y is in S_j. Then replacing x_j with y gives a different system of distinct representatives.

25. Send Timmack to Los Angeles, Alfors to Seattle, Tang to London, Ramirez to Frankfurt, Washington to Paris, Jelinek to Madrid, and Rupp to Dublin.

27. The proof is by induction on n. It is clear for $n = 1$. Let sets $S_1, S_2, \ldots, S_{k+1}$ be given, with $|S_i| \geq i$ for $i = 1, 2, \ldots, k + 1$. By the induction hypothesis, $S_1, S_2, \ldots, S_k$ has a system of distinct representatives $x_1, \ldots, x_k$. But $|S_{k+1}| > k$, so S_{k+1} contains an element x_{k+1} different from all of $x_1, \ldots, x_k$. Then $x_1, \ldots, x_k, x_{k+1}$ is a system of distinct representatives for $S_1, \ldots, S_{k+1}$. Thus the result holds for all positive integers n by the principle of mathematical induction.

121

29. Note that if $x_1, \ldots, x_k$ is a system of distinct representatives for $S_1, \ldots, S_k$, then $|S_{k+1} - \{x_1, \ldots, x_k\}| = (k+2) - k = 2$; and so there are 2 choices for x_{k+1}. Thus there are 2^n systems of distinct representatives.

31. Note that if a sequence of $n = 1$ sets S_1 satisfies Hall's condition, then $|S_1| \geq 1$, and so the sequence has a system of distinct representatives. Now assume any sequence of k or fewer sets satisfying Hall's condition has a system of distinct representatives. Suppose $S_1, \ldots, S_{k+1}$ satisfies Hall's condition.

Case 1: Whenever I is a nonempty subset of $\{1, 2, \ldots, k+1\}$ with fewer than $k+1$ elements, then $|\bigcup_{j \in I} S_i| \geq |I| + 1$.

Since $|S_{k+1}| \geq 1$, we can choose $x_{k+1} \in S_{k+1}$. Let $S_i^* = S_i - \{x_{k+1}\}$ for $i = 1, \ldots, k$. By the assumption in this case, if I is a subset of $\{1, \ldots, k\}$, then $|\bigcup_{i \in I} S_i^*| \geq |I|$, since at most one element has been taken out. Thus by the induction hypothesis, $S_1^*, \ldots, S_k^*$ has a system of distinct representatives $x_1, \ldots, x_k$. But then $x_1, \ldots, x_k, x_{k+1}$ is a system of distinct representatives for $S_1, \ldots, S_{k+1}$.

Case 2: There exists a nonempty subset I of $\{1, \ldots, k+1\}$ with fewer than $k+1$ elements such that $|\bigcup_{i \in I} S_i| = |I|$.

For simplicity suppose $I = \{1, 2, \ldots, j\}$, where $1 \leq j \leq k$. Let $A = \bigcup_{i \in I} S_i$. Then by assumption $|A| = j$. Let $S_i^* = S_i - A$ for $i = j+1, j+2, \ldots, k$. We claim that $S_{j+1}^*, \ldots, S_{k+1}^*$ satisfies Hall's condition. For example,

$$|S_{j+1}^* \cup S_{j+2}^* \cup \cdots \cup S_{j+t}^*| = |S_{j+1}^* \cup S_{j+2}^* \cup \cdots \cup S_{j+t}^*| + |A| - j$$
$$= |S_{j+1}^* \cup S_{j+2}^* \cup \cdots \cup S_{j+t}^* \cup S_1 \cup \cdots \cup S_j| - j$$
$$= |S_{j+1} \cup S_{j+2} \cup \cdots \cup S_{j+t} \cup S_1 \cup \cdots \cup S_j| - j$$
$$\geq j + t - j = t.$$

Now, by the induction hypothesis, both sequences $S_1, \ldots, S_j$ and $S_{j+1}^*, \ldots, S_{k+1}^*$ have systems of distinct representatives, say $x_1, \ldots, x_j$ and $x_{j+1}, \ldots, x_{k+1}$. But, by the definition of the sets S_i^*, these are all distinct, and so $x_1, x_2, \ldots, x_{k+1}$ is a system of distinct representatives for $S_1, \ldots, S_{k+1}$.

6.2 MATCHINGS IN GRAPHS

1. Yes; $\mathcal{V}_1 = \{1, 3, 6, 8, 9, 11, 13\}$.

3. No; if 1 is in $\mathcal{V}_1$, then 2 and 3 must be in $\mathcal{V}_2$, so 4 and 5 are in $\mathcal{V}_1$. But there is an edge between 4 and 5.

5. No; if 1 is in $\mathcal{V}_1$, then so must be 7, 8, and 2. But there is an edge between 1 and 2.

7. $\{\{1,2\}, \{3,4\}, \{5,6\}, \{7,8\}, \{9,10\}, \{12,13\}\}, \{\{1,4\}, \{3,5\}, \{6,7\}\}, \{\{1,2\}, \{3,4\}\}$

9. $\{2,4,5,7,10,12\}, \{1,3,6,7\}, \{1,2,4\}$

11.

	2	4	6	8	10	12
1	1	0	0	1	0	0
3	1	0	1	0	0	0
5	0	1	1	0	0	1
7	1	0	1	1	0	0
9	0	0	0	1	1	0
11	0	0	1	0	0	0

13.

	2	4	6	8	10	12
1	1	1	1	0	0	0
3	1	1	0	1	0	0
5	0	1	1	0	1	0
7	1	0	0	1	0	1
9	0	0	0	1	1	1
11	0	0	1	0	1	1

15.

	2	4	6	8	10	12
1	1	1	0	0	0	0
3	0	0	1	0	0	0
5	1	1	0	1	0	0
7	0	0	0	0	1	0
9	0	1	0	0	0	1
11	0	0	1	0	0	0
13	0	0	0	0	1	0

17.

$$
\begin{bmatrix}
* & 0 & 0 & 0 & 0 & 0 \\
0 & 0 & * & 0 & 0 & 0 \\
0 & * & 0 & 0 & 0 & 0 \\
0 & 0 & 0 & * & 0 & 0 \\
0 & 0 & 0 & 0 & * & 0 \\
0 & 0 & 0 & 0 & 0 & 0
\end{bmatrix},
\begin{bmatrix}
0 & 0 & 0 & * \\
* & 0 & 0 & 0 \\
0 & * & 0 & 0 \\
0 & 0 & 0 & 0 \\
0 & 0 & 0 & 0
\end{bmatrix},
\begin{bmatrix}
* & 0 & 0 & 0 & 0 & 0 \\
0 & * & 0 & 0 & 0 & 0 \\
0 & 0 & * & 0 & 0 & 0 \\
0 & 0 & 0 & * & 0 & 0 \\
0 & 0 & 0 & 0 & * & 0 \\
0 & 0 & 0 & 0 & 0 & *
\end{bmatrix}
$$

19. Row named 5, columns named 2, 6, 8, 10; rows named 3, 5, column named 8; all rows.

21. The maximum matching in the graph is indicated by the thick line segments.

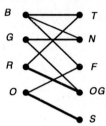

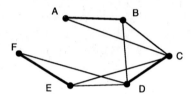

$$
\begin{array}{c}
 \\
B \\
G \\
R \\
O
\end{array}
\begin{array}{c}
T \quad N \quad F \quad OG \quad S \\
\left[
\begin{array}{ccccc}
1^* & 1 & 1 & 0 & 0 \\
0 & 1^* & 0 & 1 & 0 \\
1 & 0 & 0 & 1^* & 0 \\
0 & 0 & 1 & 0 & 1^*
\end{array}
\right]
\end{array}
$$

23. The maximum matching is indicated by by the thick line segments in the graph below.

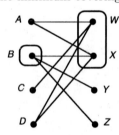

25. The minimum covering is indicated by the ovals.

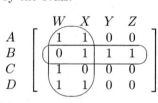

$$
\begin{array}{c}
A \\
B \\
C \\
D
\end{array}
\begin{array}{c}
W \quad X \quad Y \quad Z \\
\left[
\begin{array}{cccc}
1 & 1 & 0 & 0 \\
0 & 1 & 1 & 1 \\
1 & 0 & 0 & 0 \\
1 & 1 & 0 & 0
\end{array}
\right]
\end{array}
$$

27. Suppose that the vertices of the graph are $v_1, v_2, \ldots, v_n$, where $v_1, \ldots, v_m$ are in $\mathcal{V}_1$ and $v_{m+1}, \ldots, v_n$ are in $\mathcal{V}_2$. Then the matrix of the bipartite graph is formed by removing the last $n - m$ rows and first m columns from the adjacency matrix of the graph.

29. Let the cycle be $v_1, v_2, \ldots, v_n$, where n is odd. Suppose the graph is bipartite. If v_1 is in $\mathcal{V}_1$, then v_2 is in $\mathcal{V}_2$, v_3 is in $\mathcal{V}_1$, etc. Thus v_n is in $\mathcal{V}_1$. But this contradicts the fact that v_1 and v_n are adjacent.

31. If the graph is bipartite with vertex sets $\mathcal{V}_1$ and $\mathcal{V}_2$, color the vertices of $\mathcal{V}_1$ with one color and those of $\mathcal{V}_2$ with the other. Conversely, if the graph can be colored with red and blue, let $\mathcal{V}_1$ be the set of red vertices and $\mathcal{V}_2$ the set of blue vertices. Then every edge joins a vertex of $\mathcal{V}_1$ with one of $\mathcal{V}_2$, and the graph is bipartite.

33. For simplicity, assume $\mathcal{V} = \{1, 2, \ldots, n\}$. First assume $\mathcal{C}$ is a covering. If $a_{ij} = 1$, then $\{i, j\}$ is an edge, so $i \in \mathcal{C}$ or $j \in \mathcal{C}$ by the definition of a covering. Conversely assume $a_{ij} = 0$ whenever $i \notin \mathcal{C}$ and $j \notin \mathcal{C}$. Let $\{i, j\}$ be an edge. Then $a_{ij} = 1$, so by assumption $i \in \mathcal{C}$ or $j \in \mathcal{C}$. This proves that $\mathcal{C}$ is a covering.

6.3   A MATCHING ALGORITHM

1. Scanning row 1, we put a 1 below column B and a $\sqrt{}$ after row 1. Scanning row 2, we put a 2 below column A and a $\sqrt{}$ after row 2.

3. Circle the 1s in row 3, column A, in column A, row 2, and in row 2, column D.

5.

	A	B	C	D	
1	0	(1*)	0	(1)	D$\sqrt{}$
2	1*	1	0	0	B$\sqrt{}$
3	0	0	1*	1	D$\sqrt{}$
4	1	(1)	1	0	B!
	2	1$\sqrt{}$	3$\sqrt{}$	#$\sqrt{}$	

	A	B	C	D
1	0	1	0	1*
2	1*	1	0	0
3	0	0	1*	1
4	1	1*	1	0

7.

	A	B	C	D	E	
1	(1*)	1	0	(1)	1	D$\sqrt{}$
2	1	0	0	0	1*	A$\sqrt{}$
3	0	1*	0	1	0	D$\sqrt{}$
4	(1)	1	0	0	1	A!
	1$\sqrt{}$	3$\sqrt{}$	#$\sqrt{}$	#$\sqrt{}$	2	

	A	B	C	D	E
1	1	1	0	1*	1
2	1	0	0	0	1*
3	0	1*	0	1	0
4	1*	1	0	0	1
			#$\sqrt{}$		

9.

	A	B	C	D	E	
1	(1*)	1	1	(1)	1	D$\sqrt{}$
2	(1)	0	0	0	0	A!
3	0	1*	0	0	0	
4	1	1	0	0	0	A
5	1	0	1*	0	1	E$\sqrt{}$
	1$\sqrt{}$		5$\sqrt{}$	#$\sqrt{}$	#$\sqrt{}$	

	A	B	C	D	E	
1	1	1	1	1*	1	E$\sqrt{}$
2	1*	0	0	0	0	
3	0	1*	0	0	0	
4	1	1	0	0	0	
5	1	0	1*	0	1	E$\sqrt{}$
		5$\sqrt{}$	1$\sqrt{}$	#$\sqrt{}$		

11.

$$
\begin{array}{c}
 & \begin{array}{cccc} A & B & C & D \end{array} \\
\begin{array}{c} 1 \\ 2 \\ 3 \\ 4 \end{array} &
\left[\begin{array}{cccc}
0 & 1^* & 0 & 1 \\
0 & 1 & 0 & 0 \\
1^* & 1 & 1 & 0 \\
0 & 1 & 0 & 1^*
\end{array}\right] C\surd \\
& \begin{array}{cccc} 3\surd & & \#\surd & \end{array}
\end{array}
$$

The given matching is a maximum matching.

13.

$$
\begin{array}{c}
 & \begin{array}{ccc} A & B & C \end{array} \\
\begin{array}{c} 1 \\ 2 \\ 3 \\ 4 \end{array} &
\left[\begin{array}{ccc}
0 & 1^* & 1 \\
①^* & 0 & ① \\
① & 1 & 0 \\
0 & 1 & 0
\end{array}\right]
\begin{array}{l} C\surd \\ C\surd \\ A! \\ B \end{array} \\
& \begin{array}{ccc} 2\surd & 1\surd & \#\surd \end{array}
\end{array}
\qquad
\begin{array}{c}
 & \begin{array}{ccc} A & B & C \end{array} \\
\begin{array}{c} 1 \\ 2 \\ 3 \\ 4 \end{array} &
\left[\begin{array}{ccc}
0 & 1^* & 1 \\
1 & 0 & 1^* \\
1^* & 1 & 0 \\
0 & 1 & 0
\end{array}\right]
\end{array}
$$

A maximum matching is $\{1, B\}, \{2, C\}, \{3, A\}$.

15.

$$
\begin{array}{c}
 & \begin{array}{ccccc} A & B & C & D & E \end{array} \\
\begin{array}{c} 1 \\ 2 \\ 3 \\ 4 \\ 5 \end{array} &
\left[\begin{array}{ccccc}
1^* & 0 & 1 & 0 & 0 \\
0 & ①^* & 0 & ① & 0 \\
1 & 0 & 0 & 0 & 1^* \\
0 & ① & ①^* & 0 & 0 \\
0 & 0 & ① & 0 & 1
\end{array}\right]
\begin{array}{l} C\surd \\ D\surd \\ \\ B\surd \\ C! \end{array} \\
& \begin{array}{ccccc} 1 & 2\surd & 4\surd & \#\surd & \end{array}
\end{array}
$$

$$
\begin{array}{c}
 & \begin{array}{ccccc} A & B & C & D & E \end{array} \\
\begin{array}{c} 1 \\ 2 \\ 3 \\ 4 \\ 5 \end{array} &
\left[\begin{array}{ccccc}
1^* & 0 & 1 & 0 & 0 \\
0 & 1 & 0 & 1^* & 0 \\
1 & 0 & 0 & 0 & 1^* \\
0 & 1^* & 1 & 0 & 0 \\
0 & 0 & 1^* & 0 & 1
\end{array}\right]
\end{array}
$$

A maximum matching is $\{1, A\}, \{2, D\}, \{3, E\}, \{4, B\}, \{5, C\}$.

17.

$$
\begin{array}{c}
 & \begin{array}{cccc} A & B & C & D \end{array} \\
\begin{array}{c} 1 \\ 2 \\ 3 \\ 4 \end{array} &
\left[\begin{array}{cccc}
0 & 1^* & 0 & 0 \\
①^* & 1 & ① & 1 \\
① & 1 & 0 & 0 \\
0 & 1 & 0 & 1^*
\end{array}\right]
\begin{array}{l} \\ C\surd \\ A! \\ \end{array} \\
& \begin{array}{cccc} 2\surd & & \#\surd & \end{array}
\end{array}
\qquad
\begin{array}{c}
 & \begin{array}{cccc} A & B & C & D \end{array} \\
\begin{array}{c} 1 \\ 2 \\ 3 \\ 4 \end{array} &
\left[\begin{array}{cccc}
0 & 1^* & 0 & 0 \\
1 & 1 & 1^* & 1 \\
1^* & 1 & 0 & 0 \\
0 & 1 & 0 & 1^*
\end{array}\right]
\end{array}
$$

A system of distinct representatives is B, C, A, D.

19.

$$
\begin{array}{c}
 \\
1 \\
2 \\
3 \\
4
\end{array}
\begin{array}{cccc}
W & X & Y & Z \\
\left[\begin{array}{cccc}
1^* & 0 & 0 & 0 \\
0 & 0 & \textcircled{1*} & \textcircled{1} \\
1 & 0 & \textcircled{1} & 0 \\
1 & 1^* & 1 & 1
\end{array}\right] \\
 4\surd \;\; 2\surd \;\; \#\surd
\end{array}
\begin{array}{c}
\\
\\
Z\surd \\
Y! \\
Z\surd
\end{array}
\qquad
\begin{array}{c}
 \\
1 \\
2 \\
3 \\
4
\end{array}
\begin{array}{cccc}
W & X & Y & Z \\
\left[\begin{array}{cccc}
1^* & 0 & 0 & 0 \\
0 & 0 & 1 & 1^* \\
1 & 0 & 1^* & 0 \\
1 & 1^* & 1 & 1
\end{array}\right]
\end{array}
$$

A system of distinct representatives is W, Z, Y, X.

21.

$$
\begin{array}{c}
 \\
1 \\
2 \\
3 \\
4
\end{array}
\begin{array}{cccccc}
a & b & c & d & e & f \\
\left[\begin{array}{cccccc}
0 & 0 & 1^* & 0 & 1 & 0 \\
\textcircled{1*} & \textcircled{1} & 0 & 1 & 0 & 1 \\
1 & 0 & 1 & 0 & 1^* & 0 \\
\textcircled{1} & 0 & 1 & 0 & 1 & 0
\end{array}\right] \\
2\surd \;\; \#\surd \#\surd \;\; 3 \;\; \#\surd
\end{array}
\begin{array}{c}
\\
\\
b\surd \\
a\surd \\
a!
\end{array}
$$

$$
\begin{array}{c}
 \\
1 \\
2 \\
3 \\
4
\end{array}
\begin{array}{cccccc}
a & b & c & d & e & f \\
\left[\begin{array}{cccccc}
0 & 0 & 1^* & 0 & 1 & 0 \\
1 & 1^* & 0 & 1 & 0 & 1 \\
1 & 0 & 1 & 0 & 1^* & 0 \\
1^* & 0 & 1 & 0 & 1 & 0
\end{array}\right] \\
2\surd \#\surd \#\surd
\end{array}
\begin{array}{c}
\\
\\
d\surd \\
\\
\end{array}
$$

A system of distinct representatives is carrot, banana, egg, apple.

23.

$$
\begin{array}{c}
 \\
1 \\
2 \\
3 \\
4 \\
5
\end{array}
\begin{array}{ccccc}
A & B & C & D & E \\
\left[\begin{array}{ccccc}
0 & 1^* & 1 & 0 & 0 \\
0 & 0 & 0 & 1^* & 1 \\
0 & 1 & 0 & 1 & 1^* \\
\textcircled{1*} & 0 & \textcircled{1} & 0 & 1D\surd \\
\textcircled{1} & 1 & 0 & 1 & 0
\end{array}\right] \\
4\surd \;\; 1\surd \;\; \#\surd 3
\end{array}
\begin{array}{c}
\\
D\surd \\
\\
C\surd \\
\\
A!
\end{array}
\qquad
\begin{array}{c}
 \\
1 \\
2 \\
3 \\
4 \\
5
\end{array}
\begin{array}{ccccc}
A & B & C & D & E \\
\left[\begin{array}{ccccc}
0 & 1^* & 1 & 0 & 0 \\
0 & 0 & 0 & 1^* & 1 \\
0 & 1 & 0 & 1 & 1^* \\
1 & 0 & 1^* & 0 & 1 \\
1^* & 1 & 0 & 1 & 0
\end{array}\right]
\end{array}
$$

A better assignment is Constantine to 1, Egmont to 2, Fungo to 3, Drury to 4, and Arabella to 5.

6.4 APPLICATIONS OF THE ALGORITHM

1.

$$
\begin{array}{c@{\quad}c}
 & \begin{array}{cccc} 1 & 2 & 3 & 4 \end{array} \\
\begin{array}{c} 1 \\ 2 \\ 3 \\ 4 \end{array}
\left[
\begin{array}{cccc}
0 & 1^* & 0 & 1 \\
1^* & 1 & 1 & 0 \\
0 & 1 & 0 & 1^* \\
0 & 0 & 0 & 1
\end{array}
\right] & 3\surd \\
\quad 2\surd \qquad \#\surd &
\end{array}
$$

3.

$$
\begin{array}{c@{\quad}c}
 & \begin{array}{ccccc} 1 & 2 & 3 & 4 & 5 \end{array} \\
\begin{array}{c} 1 \\ 2 \\ 3 \\ 4 \\ 5 \end{array}
\left[
\begin{array}{ccccc}
1^* & 0 & 1 & 0 & 0 \\
0 & 0 & 1^* & 1 & 0 \\
1 & 1^* & 0 & 1 & 1 \\
1 & 0 & 1 & 1^* & 0 \\
1 & 0 & 0 & 1 & 0
\end{array}
\right] & 5\surd \\
\quad 3\surd \qquad\qquad \#\surd &
\end{array}
$$

5.

$$
\begin{array}{c@{\quad}c}
 & \begin{array}{cccc} A & B & C & D \end{array} \\
\begin{array}{c} 1 \\ 2 \\ 3 \\ 4 \end{array}
\left[
\begin{array}{cccc}
1^* & 0 & 0 & 0 \\
0 & 1^* & 1 & 1 \\
1 & 0 & 0 & 0 \\
0 & 0 & 1^* & 0
\end{array}
\right] & D\surd \\
\quad 2\surd \qquad \#\surd &
\end{array}
$$

7.

$$
\begin{array}{c@{\quad}c@{\qquad}c}
\begin{array}{c} \\ 1 \\ 2 \\ 3 \\ 4 \\ 5 \\ \end{array}
\begin{array}{c}
\begin{array}{ccccc} A & B & C & D & E \end{array} \\
\left[
\begin{array}{ccccc}
0 & 1^* & 0 & 0 & 0 \\
0 & 0 & 0 & 1^* & 0 \\
0 & 1 & 1^* & 0 & 1 \\
0 & 1 & 1 & 1 & 0 \\
0 & 1 & 0 & 1 & 0
\end{array}
\right] \\
\#\surd \ \ 3\surd \ \ \#\surd
\end{array}
&
\begin{array}{c} E\surd \\ C! \end{array}
&
\begin{array}{c}
\begin{array}{c} \\ 1 \\ 2 \\ 3 \\ 4 \\ 5 \\ \end{array}
\end{array}
\begin{array}{c}
\begin{array}{ccccc} A & B & C & D & E \end{array} \\
\left[
\begin{array}{ccccc}
0 & 1^* & 0 & 0 & 0 \\
0 & 0 & 0 & 1^* & 0 \\
0 & 1 & 1 & 0 & 1^* \\
0 & 1 & 1^* & 1 & 0 \\
0 & 1 & 0 & 1 & 0
\end{array}
\right] \\
\#\surd
\end{array}
\end{array}
$$

The unlabeled columns in the second matrix give the covering $\{B, C, D, E\}$.

9.

$$
\begin{array}{c@{\quad}c}
 & \begin{array}{ccccc} 1 & 2 & 3 & 4 & 5 \end{array} \\
\begin{array}{c} 1 \\ 2 \\ 3 \\ 4 \\ 5 \end{array}
\left[
\begin{array}{ccccc}
0 & 1^* & 0 & 1 & 1 \\
1^* & 0 & 1 & 0 & 1 \\
0 & 1 & 1^* & 0 & 1 \\
0 & 0 & 1 & 1^* & 1 \\
0 & 1 & 1 & 1 & 0
\end{array}
\right] &
\begin{array}{c} 5\surd \\ 5\surd \\ 5\surd \\ 5\surd \\ 2! \end{array} \\
2\surd \ 1\surd \ 3\surd \ 4\surd \ \#\surd &
\end{array}
$$

There is no such set I, since a system of distinct representatives exists.

11.

$$
\begin{array}{c}
 \\
1 \\
2 \\
3 \\
4 \\
5 \\
6
\end{array}
\begin{array}{c}
\begin{array}{ccccccc}
1 & 2 & 3 & 4 & 5 & 6 & 7
\end{array} \\
\left[
\begin{array}{ccccccc}
0 & 1^* & 0 & 0 & 0 & 0 & 1 \\
1^* & 0 & 1 & 0 & 0 & 1 & 0 \\
0 & 0 & 0 & 0 & 1^* & 0 & 1 \\
0 & 0 & 1^* & 1 & 0 & 1 & 0 \\
0 & 1 & 0 & 0 & 1 & 0 & 0 \\
0 & 1 & 0 & 0 & 1 & 0 & 1^*
\end{array}
\right] \\
\begin{array}{ccccccc}
2\surd & & 4\surd & \#\surd & & \#\surd &
\end{array}
\end{array}
\quad
\begin{array}{l}
\\
\\
6\surd \\
\\
4\surd \\
\\
\\
\end{array}
$$

Using the unlabeled rows, we take $I = \{1, 3, 5, 6\}$.

13. The matrix giving the times of runners to posts is

$$
\begin{array}{c}
\\
1 \\
2 \\
3 \\
4
\end{array}
\begin{array}{c}
\begin{array}{cccc}
A & B & C & D
\end{array} \\
\left[
\begin{array}{cccc}
6 & 4 & 5 & 7 \\
5 & 8 & 3 & 6 \\
9 & 7 & 9 & 3 \\
7 & 8 & 8 & 5
\end{array}
\right]
\end{array}.
$$

Since 4 runners are needed, it will take at least 5 hours. Using runners with times no more than 5 hours gives the matrix

$$
\begin{bmatrix}
0 & 1 & 1 & 0 \\
1 & 0 & 1 & 0 \\
0 & 0 & 0 & 1 \\
0 & 0 & 0 & 1
\end{bmatrix}.
$$

Considering rows 3 and 4 and the last column, we see that no independent set with 4 entries exists, and the same is true if we allow 6 hours. Allowing 7 hours gives the matrix

$$
\begin{bmatrix}
1^* & 1 & 1 & 1 \\
1 & 0 & 1^* & 1 \\
0 & 1^* & 0 & 1 \\
1 & 0 & 0 & 1^*
\end{bmatrix},
$$

which has the independent set indicated.

15. We need a minimum covering for the matrix

$$
\begin{array}{c}
\\
1 \\
2 \\
3 \\
4 \\
5 \\
6 \\
7 \\
8
\end{array}
\begin{array}{c}
\begin{array}{ccccccccc}
A & B & C & D & E & F & G & H & I
\end{array} \\
\left[
\begin{array}{ccccccccc}
1^* & 0 & 0 & 0 & 1 & 0 & 0 & 0 & 0 \\
0 & 1^* & 0 & 0 & 0 & 0 & 0 & 0 & 0 \\
0 & 1 & 0 & 0 & 0 & 0 & 0 & 0 & 0 \\
0 & 1 & 1^* & 1 & 0 & 0 & 1 & 0 & 0 \\
0 & 0 & 0 & 1^* & 1 & 1 & 0 & 0 & 0 \\
0 & 0 & 0 & 0 & 0 & 1^* & 0 & 0 & 1 \\
0 & 0 & 1 & 0 & 0 & 0 & 0 & 1^* & 0 \\
0 & 0 & 0 & 0 & 0 & 0 & 1^* & 1 & 1
\end{array}
\right] \\
\begin{array}{ccccccccc}
1\surd & & 4\surd & 5\surd & \#\surd & 6\surd & 8\surd & 7\surd & \#\surd
\end{array}
\end{array}
\quad
\begin{array}{l}
E\surd \\
\\
\\
D\surd \\
E\surd \\
I\surd \\
C\surd \\
I\surd
\end{array}
$$

129

The independent set above was found by inspection. The algorithm indicates it is maximum. Thus the labeled rows and unlabeled columns give the minimum covering $\{1, 4, 5, 6, 7, 8, B\}$.

17. Let S be the union of the sets corresponding to unlabeled rows. Each element of S corresponds to a 1 in an unlabeled row. If this 1 has no star, then its column is unlabeled, since otherwise the row would be. If it has a star, then the column is also unlabeled since each column has at most one star. Thus $|S| \le c_U$. Now consider an unlabeled column. It must contain a 1*, since otherwise it would be labeled at step 1. This 1* must be in an unlabeled row, otherwise the column would be labeled. Thus $c_U \le |S|$.

6.5 THE HUNGARIAN METHOD

1.

$$
\begin{bmatrix} 1 & 2 & 3 \\ 6 & 5 & 4 \\ 7 & 8 & 9 \end{bmatrix}
\quad
\begin{bmatrix} 0 & 1 & 2 \\ 2 & 1 & 0 \\ 0 & 1 & 2 \end{bmatrix}
\quad
\begin{bmatrix} 0^* & 0 & 2 \\ 2 & 0 & 0^* \\ 0 & 0^* & 2 \end{bmatrix}
$$

We find the indicated independent set of 0s by inspection. The sum of the corresponding entries in the original matrix is $1 + 4 + 8 = 13$.

3.

$$
\begin{bmatrix} 6 & 2 & 5 & 8 \\ 6 & 7 & 1 & 6 \\ 6 & 3 & 4 & 5 \\ 5 & 4 & 3 & 4 \end{bmatrix}
\quad
\begin{bmatrix} 4 & 0 & 3 & 6 \\ 5 & 6 & 0 & 5 \\ 3 & 0 & 1 & 2 \\ 2 & 1 & 0 & 1 \end{bmatrix}
\quad
\begin{bmatrix} 2 & 0 & 3 & 5 \\ 3 & 6 & 0 & 4 \\ 1 & 0 & 1 & 1 \\ 0 & 1 & 0 & 0 \end{bmatrix}
$$

By inspection the 3 lines shown contain all the 0s. The new matrix is

$$
\begin{bmatrix} 1 & 0^* & 3 & 4 \\ 2 & 6 & 0^* & 3 \\ 0^* & 0 & 1 & 0 \\ 0 & 2 & 1 & 0^* \end{bmatrix}
$$

The independent set of 0s shown corresponds to the sum $2 + 1 + 6 + 4 = 13$.

5.

$$
\begin{bmatrix} 3 & 5 & 5 & 3 & 8 \\ 4 & 6 & 4 & 2 & 6 \\ 4 & 6 & 1 & 3 & 6 \\ 3 & 4 & 4 & 6 & 5 \\ 5 & 7 & 3 & 5 & 9 \end{bmatrix}
\quad
\begin{bmatrix} 0 & 2 & 2 & 0 & 5 \\ 2 & 4 & 2 & 0 & 4 \\ 3 & 5 & 0 & 2 & 5 \\ 0 & 1 & 1 & 3 & 2 \\ 2 & 7 & 0 & 2 & 6 \end{bmatrix}
\quad
\begin{bmatrix} 0^* & 1 & 2 & 0 & 3 \\ 2 & 3 & 2 & 0^* & 2 \\ 3 & 4 & 0^* & 2 & 3 \\ 0 & 0^* & 1 & 3 & 0 \\ 2 & 3 & 0 & 2 & 4 \end{bmatrix}
$$

The independent set algorithm applied to the last matrix shows that all the 0s are in row 4 and columns 1, 3, and 4. The Hungarian method produces the matrix

$$\begin{bmatrix} 0^* & 0 & 2 & 0 & 2 \\ 2 & 2 & 2 & 0^* & 1 \\ 3 & 3 & 0^* & 2 & 2 \\ 1 & 0^* & 2 & 4 & 0 \\ 2 & 2 & 0 & 2 & 3 \end{bmatrix}$$

Now all the 0s are in rows 1 and 4 and columns 3 and 4. Using the Hungarian algorithm again gives

$$\begin{bmatrix} 0^* & 0 & 3 & 1 & 2 \\ 1 & 1 & 2 & 0^* & 0 \\ 2 & 2 & 0^* & 2 & 1 \\ 1 & 0^* & 3 & 5 & 0 \\ 1 & 1 & 0 & 2 & 2 \end{bmatrix}$$

Now all the 0s are in rows 1, 2, and 4 and column 3. The next matrix is

$$\begin{bmatrix} 0^* & 0 & 4 & 1 & 2 \\ 1 & 1 & 3 & 0^* & 0 \\ 1 & 1 & 0^* & 1 & 0 \\ 1 & 0 & 4 & 5 & 0^* \\ 0 & 0^* & 0 & 1 & 1 \end{bmatrix}$$

The indicated independent set corresponds to the sum $3 + 2 + 1 + 5 + 7 = 18$.

7.

$$\begin{bmatrix} 3 & 4 & 5 & 7 & 6 \\ 5 & 3 & 4 & 5 & 2 \\ 1 & 3 & 4 & 5 & 3 \\ 5 & 6 & 5 & 4 & 3 \\ 0 & 0 & 0 & 0 & 0 \end{bmatrix} \quad \begin{bmatrix} 0 & 1 & 2 & 4 & 3 \\ 3 & 1 & 2 & 3 & 0 \\ 0 & 2 & 3 & 4 & 2 \\ 2 & 3 & 2 & 1 & 0 \\ 0 & 0 & 0 & 0 & 0 \end{bmatrix} \quad \begin{bmatrix} 0 & 0^* & 1 & 3 & 3 \\ 3 & 0 & 1 & 2 & 0^* \\ 0^* & 1 & 2 & 3 & 2 \\ 2 & 2 & 1 & 0^* & 0 \\ 1 & 0 & 0^* & 0 & 1 \end{bmatrix}$$

The independent set shown corresponds to the sum $4 + 2 + 1 + 4 + 0 = 11$.

9.

$$\begin{bmatrix} -5 & -4 & -2 & -3 \\ -3 & -1 & -4 & -3 \\ -1 & -1 & -1 & -3 \\ -5 & -3 & -6 & -3 \end{bmatrix} \quad \begin{bmatrix} 0 & 1 & 3 & 2 \\ 1 & 3 & 0 & 1 \\ 2 & 2 & 2 & 0 \\ 1 & 3 & 0 & 3 \end{bmatrix} \quad \begin{bmatrix} 0 & 0 & 3 & 2 \\ 1 & 2 & 0 & 1 \\ 2 & 1 & 2 & 0 \\ 1 & 2 & 0 & 3 \end{bmatrix} \quad \begin{bmatrix} 0 & 0^* & 4 & 3 \\ 0^* & 1 & 0 & 1 \\ 1 & 0 & 2 & 0^* \\ 0 & 1 & 0^* & 3 \end{bmatrix}$$

The independent set shown corresponds to the sum $4 + 3 + 3 + 6 = 16$.

131

11.

$$\begin{bmatrix} -6 & -5 & -3 & -1 & -4 \\ -2 & -5 & -3 & -7 & -8 \\ -8 & -3 & -7 & -5 & -4 \\ -7 & -1 & -5 & -3 & -8 \\ 0 & 0 & 0 & 0 & 0 \end{bmatrix} \quad \begin{bmatrix} 0 & 1 & 3 & 5 & 2 \\ 6 & 3 & 5 & 1 & 0 \\ 0 & 5 & 1 & 3 & 4 \\ 1 & 7 & 3 & 5 & 0 \\ 0 & 0 & 0 & 0 & 0 \end{bmatrix} \quad \begin{bmatrix} 0^* & 0 & 2 & 4 & 2 \\ 6 & 2 & 4 & 0^* & 0 \\ 0 & 4 & 0^* & 2 & 4 \\ 1 & 6 & 2 & 4 & 0^* \\ 1 & 0^* & 0 & 0 & 1 \end{bmatrix}$$

The independent set indicated corresponds to the sum $6 + 7 + 7 + 8 + 0 = 28$.

13. The initial matrix is

$$\begin{array}{cccc} \text{LA} & \text{NY} & \text{LV} & \text{Chi} \\ \begin{bmatrix} 700 & 500 & 200 & 400 \\ 500 & 500 & 100 & 600 \\ 500 & 300 & 400 & 700 \\ 400 & 500 & 600 & 500 \end{bmatrix} \end{array}.$$

We apply the Hungarian method to this matrix to find an assignment giving a minimum sum.

$$\begin{bmatrix} 500 & 300 & 0 & 200 \\ 400 & 400 & 0 & 500 \\ 200 & 0 & 100 & 500 \\ 0 & 100 & 200 & 100 \end{bmatrix} \quad \begin{bmatrix} 500 & 300 & 0 & 100 \\ 400 & 400 & 0 & 400 \\ 200 & 0 & 100 & 300 \\ 0 & 100 & 200 & 0 \end{bmatrix}$$

Since all 0s are in rows 3 and 4 and column 3, we obtain the matrix

$$\begin{bmatrix} 400 & 200 & 0 & 0^* \\ 300 & 300 & 0^* & 300 \\ 200 & 0^* & 200 & 300 \\ 0^* & 100 & 300 & 0 \end{bmatrix}.$$

The indicated independent set corresponds to sending Addams to Chicago, Hart to Las Vegas, Young to New York, and Herriman to Los Angeles.

15. The Hungarian algorithm can only be applied to a square matrix. In this example, independent sets with only 4 entries are not treated equally when a number is subtracted from each entry in a column, since the independent set may or may not have an entry in that column.

SUPPLEMENTARY EXERCISES

1. (a) $5 \cdot 4 \cdot 3 = 60$
   (b) $4 \cdot 3 \cdot 3 = 36$
   (c) none, since there are 5 sets but only 4 elements

3. (a) not bipartite, since 2, 3, 6, 9, 8 is an odd cycle
   (b) bipartite, with $\mathcal{V}_1 = \{1, 2, 5, 6, 7, 8, 11, 12\}$ and $\mathcal{V}_2 = \{3, 4, 9, 10, 13, 14, 15, 16\}$

5. (a) Note that the graph has 16 edges and that no vertex has degree more than 3. Thus a covering must have at least 6 vertices. One is $\{2, 4, 6, 7, 9, 11\}$.
   (b) Since the graph is bipartite, the independent set algorithm can be applied to show that a minimum covering has 8 elements. One is $\{1, 2, 5, 6, 7, 8, 11, 12\}$.

7.

$$
\begin{array}{c c}
 & \begin{array}{cccc} 2 & 4 & 5 & 7 \end{array} \\
\begin{array}{c} 1 \\ 3 \\ 6 \\ 8 \end{array} &
\left[\begin{array}{cccc}
1^* & 0 & 1 & 0 \\
1 & 1^* & 1 & 0 \\
0 & 1 & 0 & 1^* \\
0 & 0 & 1^* & 1
\end{array} \right]
\end{array}
$$

9.

$$
\begin{array}{c c c}
 & \begin{array}{ccccc} A & B & C & D & E \end{array} & \\
\begin{array}{c} 1 \\ 2 \\ 3 \\ 4 \\ 5 \end{array} &
\left[\begin{array}{ccccc}
1^* & 0 & 0 & 0 & 1 \\
0 & 0 & 1^* & 1 & 0 \\
0 & 0 & 1 & 0 & 0 \\
1 & 1^* & 0 & 1 & 1 \\
0 & 0 & 1 & 0 & 0
\end{array} \right] &
\begin{array}{c} E\surd \\ D\surd \\ C! \\ D\surd \\ C \end{array} \\
 & \begin{array}{ccccc} 1\surd & 4\surd & 2\surd & \#\surd & \#\surd \end{array} &
\end{array}
$$

$$
\begin{array}{c c c}
 & \begin{array}{ccccc} A & B & C & D & E \end{array} & \\
\begin{array}{c} 1 \\ 2 \\ 3 \\ 4 \\ 5 \end{array} &
\left[\begin{array}{ccccc}
1^* & 0 & 0 & 0 & 1 \\
0 & 0 & 1 & 1^* & 0 \\
0 & 0 & 1^* & 0 & 0 \\
1 & 1^* & 0 & 1 & 1 \\
0 & 0 & 1 & 0 & 0
\end{array} \right] &
\begin{array}{c} E\surd \\ \\ \\ E\surd \\ \end{array} \\
 & \begin{array}{ccccc} 1\surd & 4\surd & & \#\surd & \end{array} &
\end{array}
$$

A maximum independent set and minimum covering are shown in the second matrix.

11.

$$
\begin{array}{c c}
\begin{array}{c}
\\
1 \\
2 \\
3 \\
4 \\
5 \\
\end{array}
&
\begin{array}{ccccc}
v & w & x & y & z \\
0 & 1^* & 0 & 1 & 0 \\
0 & 0 & \textcircled{1}^* & 0 & \textcircled{1} \\
1^* & 0 & 0 & 0 & 1 \\
0 & 1 & \textcircled{1} & 0 & 0 \\
1 & 0 & 0 & 1^* & 0 \\
\end{array}
\end{array}
\qquad
\begin{array}{c}
\\ \\ z\surd \\ z\surd \\ x! \\ v \\
\end{array}
\qquad
\begin{array}{c}
\\
1 \\
2 \\
3 \\
4 \\
5 \\
\end{array}
\begin{array}{ccccc}
v & w & x & y & z \\
0 & 1^* & 0 & 1 & 0 \\
0 & 0 & 1 & 0 & 1^* \\
1^* & 0 & 0 & 0 & 1 \\
0 & 1 & 1^* & 0 & 0 \\
1 & 0 & 0 & 1^* & 0 \\
\end{array}
$$

$$3\surd \qquad 2\surd \qquad \#\surd$$

The system of distinct representatives is w, z, v, x, y.

13. By looking at jobs 1 and 3, we see that 4 hours is not enough. The assignments $\{1, B\}, \{2, C\}, \{3, E\}, \{4, D\}, \{5, A\}$ do the job in 5 hours.

15.

$$
\begin{array}{c}
A \\
B \\
C \\
D \\
\end{array}
\begin{array}{cccc}
\text{Hup} & \text{S} & \text{P} & \text{Hud} \\
-6 & -8 & -7 & -4 \\
-7 & -3 & -2 & -5 \\
-6 & -7 & -8 & -7 \\
-6 & -4 & -5 & -4 \\
\end{array}
\qquad
\begin{array}{cccc}
2 & 0 & 1 & 4 \\
0 & 4 & 5 & 2 \\
2 & 1 & 0 & 1 \\
0 & 2 & 1 & 2 \\
\end{array}
\qquad
\begin{array}{cccc}
2 & 0 & 1 & 3 \\
0 & 4 & 5 & 1 \\
2 & 1 & 0 & 0 \\
0 & 2 & 1 & 1 \\
\end{array}
$$

The maximum sum is $8 + 7 + 8 + 4 = 27$ cars, with Adam, Studebakers; Beth, Hupmobiles; Cal, Packards; and Danielle, Hudsons.

17. A matching with more edges can be formed by removing $e_2, e_4, \ldots, e_{n-1}$ from $\mathcal{M}$ and including $e_1, e_3, \ldots, e_n$.

Chapter 7

Network Flows

7.1 FLOWS AND CUTS

1. This weighted directed graph is a network with source A and sink E.

3. This weighted directed graph is not a network because arc (C, B) has negative capacity.

5. This weighted directed graph is a network with source D and sink B.

7. This set of numbers is not a flow because the quantity coming into D is 5 and the quantity coming out of D is 6.

9. This set of numbers is a flow with value 3, the sum of the flows along the arcs (A, B) and (A, C) leaving the source.

11. This set of numbers is not a flow because the quantity coming into D is 2 and the quantity coming out of D is 3.

13. These sets do not form a cut because vertex C is not in $\mathcal{S}$ or $\mathcal{T}$.

15. These sets form a cut with capacity 40, the sum of the capacities along the arcs (A, B), (A, C), (D, F), (E, C), and (E, F).

17. These sets form a cut with capacity 34, the sum of the capacities along the arcs (A, B), (A, C), (D, F), and (E, F).

19. The diagram below shows a flow with value 11.

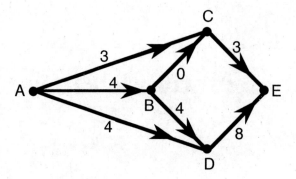

21. The diagram below shows a flow with value 11.

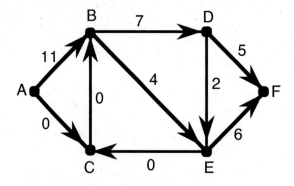

23. The diagram below shows a flow with value 18.

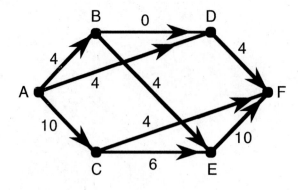

25. The sets $\{A, B, C, D\}$, $\{E\}$ form a cut with capacity 11, the sum of the capacities along the arcs (C, E) and (D, E).

27. The sets $\{A, B, C\}$, $\{D, E, F\}$ form a cut with capacity 11, the sum of the capacities along the arcs (B, D) and (B, E).

29. The sets $\{A, C\}$, $\{B, D, E, F\}$ form a cut with capacity 18, the sum of the capacities along the arcs (A, B), (A, D), (C, E), and (C, F).

31. The diagram below shows the desired network.

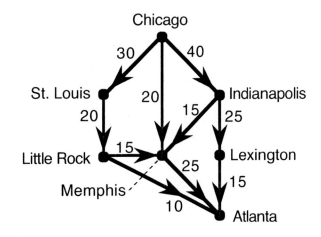

33. For the flow in Exercise 10, we have

$$f(\mathcal{U}, \mathcal{V}) = f(B, E) + f(C, A) + f(D, E) + f(D, F) = 5 + 1 + 4 + 5 = 15$$

and

$$f(\mathcal{V}, \mathcal{U}) = f(A, B) + f(E, C) = 1 + 2 = 3.$$

35. For the flow in Exercise 10, take $\mathcal{U} = \{D\}$, $\mathcal{V}_1 = \{A, B, C\}$, and $\mathcal{V}_2 = \{B, E, F\}$. Then

$$f(\mathcal{U}, \mathcal{V}_1 \cup \mathcal{V}_2) = f(D, B) + f(D, E) + f(D, F) = 3 + 4 + 5 = 12,$$

but

$$f(\mathcal{U}, \mathcal{V}_1) = f(D, B) = 3 \quad \text{and} \quad f(\mathcal{U}, \mathcal{V}_2) = f(D, B) + f(D, E) + f(D, F) = 3 + 4 + 5 = 12.$$

7.2   A FLOW AUGMENTATION ALGORITHM

1. The flow can be increased by 1 along the directed path A, B, D, E because the flow along arc (B, D) can only be increased from 2 to 3.

3. The flow can be increased by 2 along the directed path A, B, E, D, F because the flow along arc (D, E) can only be decreased from 2 to 0.

5. A flow with a larger value is obtained by increasing the flow by 4 along the directed path A, D, E. This flow is shown in the diagram below.

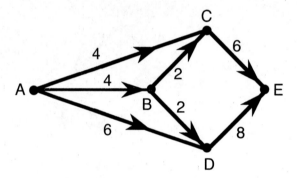

7. A flow with a larger value is obtained by increasing the flow by 2 along the directed path A, B, E, D, F. This flow is shown in the diagram below.

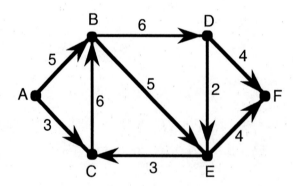

138

9. When the flow augmentation algorithm is applied to the given network, the resulting labels are shown in the diagram below.

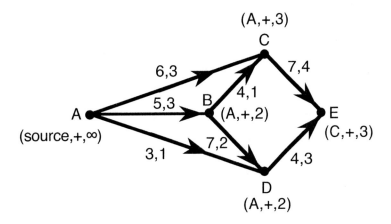

Because the sink is labeled $(C, +, 3)$, the flow can be increased by 3. This label also indicates that the vertex that precedes E on the flow-augmenting path is C, the first entry of the label assigned to E. Similarly, the label $(A, +, 3)$ assigned to C shows that the vertex that precedes C on the flow-augmenting path is A. Hence the flow-augmenting path is A, C, E.

11. When the flow augmentation algorithm is applied to the given network, the resulting labels are shown in the diagram below.

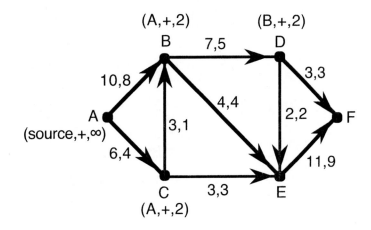

Since only vertices A, B, C, and D can be labeled, the given flow is maximal.

13. When the flow augmentation algorithm is applied to the given network, the resulting labels are shown in the diagram below.

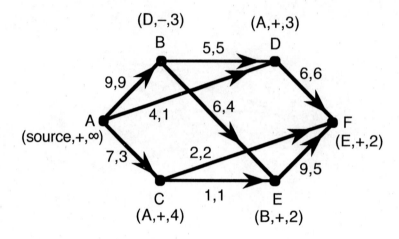

Since the sink is labeled, the flow can be increased. The labels assigned to the vertices show that the flow can be increased by 2 along the directed path A, D, B, E, F.

15. When the flow augmentation algorithm is applied to the given network, the resulting labels are shown in the diagram below.

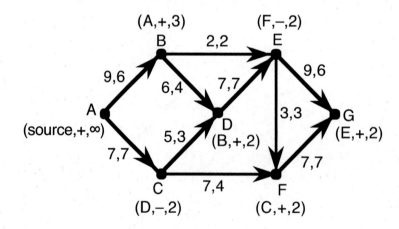

Thus the flow can be increased by 2 along the directed path A, B, D, C, F, E, G.

17. Since only vertices A and C can be labeled, the given flow is maximal.

19. The flow can be increased by 4 along the directed path A, B, E, F, G. The resulting flow, shown below, is maximal because the flow along each arc leaving A is equal to the arc's capacity.

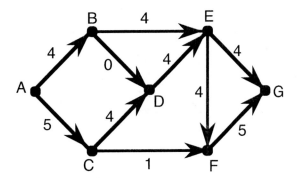

21. In the first iteration of the flow augmentation algorithm, 2 units will be sent along the directed path A, C, E. In the second iteration, 3 units will be sent along A, D, E. In the third, 2 units will be sent along A, B, C, E. In the fourth iteration, 2 units will be sent along A, B, D, E. The resulting maximal flow is shown below.

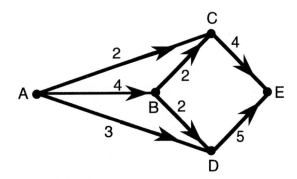

23. In the first iteration of the flow augmentation algorithm, 45 units will be sent along the directed path A, C, F. In the second iteration, 50 units will be sent along A, D, F. In the third iteration, 15 units will be sent along A, B, D, F. In the fourth iteration, 55 units will be sent along A, B, E, F. The resulting maximal flow is shown on the next page.

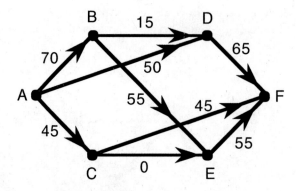

25. In the first iteration of the flow augmentation algorithm, 3 units will be sent along the directed path A, B, D, F. In the second iteration, 2 units will be sent along A, C, D, B, E, F. The resulting maximal flow is shown below.

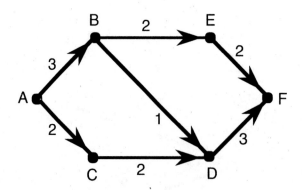

27. In the first iteration of the flow augmentation algorithm, 3 units will be sent along the directed path A, D, G. In the second iteration, 3 units will be sent along A, B, E, G. In the third iteration, 4 units will be sent along A, C, F, G. In the fourth iteration, 4 units will be sent along A, D, E, G. In the fifth iteration, 2 units will be sent along A, D, F, G. In the sixth iteration, 1 unit will be sent along A, B, D, F, G. In the seventh iteration, 1 unit will be sent along A, C, D, F, G. The resulting maximal flow has a value of 18, and a minimal cut is $\{A, B, C, D\}, \{E, F, G\}$. The maximal flow is shown on the next page.

142

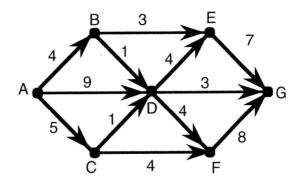

29. One possible example is shown below.

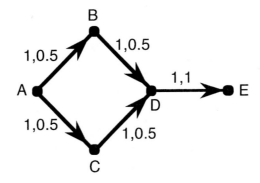

7.3 THE MAX-FLOW MIN-CUT THEOREM

1. The capacity of the cut is

$$c(A, B) + c(C, D) + c(F, G) = 9 + 5 + 7 = 21.$$

3. The capacity of the cut is

$$c(A, B) + c(A, C) + c(E, F) + c(E, G) = 9 + 7 + 3 + 9 = 28.$$

5. When the flow augmentation algorithm is applied to the given network, only the vertices A, B, C, and E receive labels. Thus $\{A, B, C, E\}, \{D, F\}$ is a minimal cut.

7. When the flow augmentation algorithm is applied to the given network, only the vertices A, B, and D receive labels. Thus $\{A, B, D\}, \{C, E, F\}$ is a minimal cut.

143

9. By applying the flow augmentation algorithm to a maximal flow, we see as in Exercise 5 that $\{A, B, C, D\}$, $\{E\}$ is a minimal cut .

11. As in Exercise 9, a minimal cut is $\{A, C, F\}$, $\{B, D, E, G\}$.

13. Multiply all the capacities by 12 to obtain the network $\mathcal{N}'$. When the flow augmentation algorithm is applied to $\mathcal{N}'$, 14 units are sent along the path A, C, E in the first iteration and 18 units are sent along the path A, D, E in the second iteration. The corresponding maximal flow in the original network is shown below.

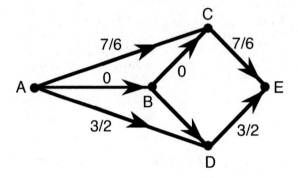

15. (a) Since the arc capacities of $\mathcal{N}'$ are obtained by multiplying the arc capacities of $\mathcal{N}$ by d, the capacity of a cut $\mathcal{S}, \mathcal{T}$ in $\mathcal{N}'$ is d times the capacity of that cut in $\mathcal{N}$. Hence $\mathcal{S}, \mathcal{T}$ is a minimal cut in $\mathcal{N}'$ if and only if it is a minimal cut in $\mathcal{N}$.

(b) Let v and v' be the values of maximal flows for $\mathcal{N}$ and $\mathcal{N}'$, respectively. Then the capacities of minimal cuts for N and $\mathcal{N}'$ are also v and v' by the max-flow min-cut theorem. Hence by (a), $v' = dv$.

(c) Define f' by $f'(x, y) = d \cdot f(x, y)$. Clearly f is a flow in N if and only if f' is a flow in N', and in this case the value of f' is d times the value of f. Thus, by (b), f is a maximal flow for N if and only if f' is a maximal flow for N'.

17. Consider a transportation network with vertices $v_1, v_2, \ldots, v_n$, where v_1 is the source and v_n is the sink. If $\mathcal{S}, \mathcal{T}$ is a cut, then $v_1 \in \mathcal{S}$, and $v_n \notin \mathcal{S}$. Thus $\mathcal{S} = \{v_1\} \cup A$, where A is a subset of $\{v_2, v_3, \ldots, v_{n-1}\}$. Moreover, every such set A determines a cut. Thus the number of cuts is the same as the number of subsets of $\{v_2, v_3, \ldots, v_{n-1}\}$, which is 2^{n-2}.

19. Make the given directed graph into a network by giving each arc a capacity of 1. By applying the flow augmentation algorithm, we find that a maximal flow has value 2. Hence, by Exercise 18, the minimal number of directed edges whose removal leaves no directed path from s to t is 2. One such set is $\{(S, A), (F, T)\}$.

21. By replacing each edge $\{x, y\}$ of $\mathcal{G}$ by two directed edges (x, y) and (y, x) with capacity $c(x, y)$, we obtain a weighted directed graph $\mathcal{N}$ of the type described in Exercise 16.

That exercise implies that a maximal flow in $\mathcal{G}$ corresponds to a maximal flow in $\mathcal{N}$. Since the flow augmentation algorithm produces a maximal flow in this directed graph, it also gives a maximal flow in $\mathcal{G}$.

23. When the flow augmentation algorithm is applied as described in Exercise 21, 4 units are sent along the path S, A, D, T in the first iteration, 3 units are sent along the path S, B, E, T in the second iteration, 2 units are sent along the path S, A, C, D, T in the third iteration, and 1 unit is sent along S, A, B, C, D, T in the fourth iteration. The resulting maximal flow is shown below.

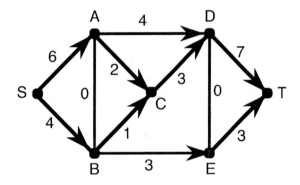

25. Let n denote the number of vertices in $\mathcal{V}$. The sum z of the capacities of every cut for the transportation network is a sum of terms of the form $ac(X, Y)$, where $X, Y \in \mathcal{V}$ and a is the number of cuts $\mathcal{S}, \mathcal{T}$ with $X \in \mathcal{S}$ and $Y \in \mathcal{T}$. We consider four cases.

Case 1: $X = S$ and $Y = T$.

Since, by Exercise 17, there are 2^{n-2} possible cuts and each cut capacity contains the term $c(S, T)$, we see that z contains the term $2^{n-2}c(S, T)$.

Case 2: $X = S$ and $Y \neq T$.

If in cut $\mathcal{S}, \mathcal{T}$ we have $Y \in \mathcal{S}$, then $c(S, Y)$ is not a term in the capacity of cut $\mathcal{S}, \mathcal{T}$. This occurs in half of the cuts in the network. In the other half of the cuts $\mathcal{S}, \mathcal{T}$, we have $Y \in \mathcal{T}$, and $c(S, Y)$ is a term in the capacity of cut $\mathcal{S}, \mathcal{T}$. Thus z contains a term of the form $2^{n-3}c(S, Y)$.

Case 3: $X \neq S$ and $Y = T$.

As in Case 2, z contains a term of the form $2^{n-3}c(X, T)$.

Case 4: $X \neq S$ and $Y \neq T$.

A term $c(X, Y)$ does not affect the capacity of cut $\mathcal{S}, \mathcal{T}$ unless $X \in \mathcal{S}$ and $Y \in \mathcal{T}$. Because $X \in \mathcal{S}$ and $Y \in \mathcal{T}$ in one-fourth of all the cuts $\mathcal{S}, \mathcal{T}$, it follows that z contains a term of the form $2^{n-4}c(X, Y)$.

Define

$$
p = \sum_{\substack{Y \in \mathcal{V} \\ Y \neq T}} c(S, Y), \quad q = \sum_{\substack{X \in \mathcal{V} \\ X \neq S}} c(X, T), \quad \text{and} \quad r = \sum_{\substack{X, Y \in \mathcal{V} \\ X \neq S \\ Y \neq T}} c(X, Y).
$$

Then

$$
\begin{aligned}
z &= 2^{n-2}c(S, T) + 2^{n-3}p + 2^{n-3}q + 2^{n-4}r \\
&= 4 \cdot 2^{n-4}c(S, T) + 2 \cdot 2^{n-4}p + 2 \cdot 2^{n-4}q + 2^{n-4}r \\
&= 3 \cdot 2^{n-4}c(S, T) + 2^{n-4}p + 2^{n-4}q + 2^{n-4} \cdot \sum_{X, Y \in \mathcal{V}} c(X, Y) \\
&= 2^{n-4}c(S, T) + 2^{n-4} \cdot \sum_{Y \in \mathcal{V}} c(S, Y) + 2^{n-4} \cdot \sum_{X \in \mathcal{V}} c(X, T) + 2^{n-4} \cdot \sum_{X, Y \in \mathcal{V}} c(X, Y) \\
&= 2^{n-4} \left(c(S, T) + \sum_{Y \in \mathcal{V}} c(S, Y) + \sum_{X \in \mathcal{V}} c(X, T) + \sum_{X, Y \in \mathcal{V}} c(X, Y) \right).
\end{aligned}
$$

Since there are 2^{n-2} possible cuts, the average capacity of a cut is

$$
\frac{z}{2^{n-2}} = \frac{1}{4} \left(c(S, T) + \sum_{Y \in \mathcal{V}} C(S, Y) + \sum_{X \in \mathcal{V}} C(X, T) + \sum_{X, Y \in \mathcal{V}} C(X, Y) \right).
$$

7.4   FLOWS AND MATCHINGS

1. The graph is bipartite with $\mathcal{V}_1 = \{A, D, E\}$ and $\mathcal{V}_2 = \{B, C, F\}$. The network associated with the given graph is shown on the next page. (In this network all arcs have capacity 1 and are directed from left to right.)

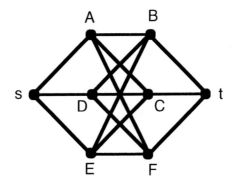

3. The graph is not bipartite because A, B, F, E, C, A is a cycle of odd length.

5. The graph is bipartite with $\mathcal{V}_1 = \{A, D\}$ and $\mathcal{V}_2 = \{B, C, E, F\}$. The network associated with the given graph is shown below. (In this network all arcs have capacity 1 and are directed from left to right.)

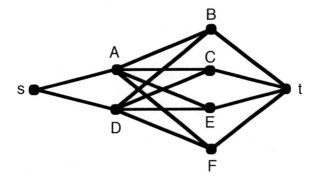

7. The flow augmentation algorithm gives the matching $\{(A, Y), (B, Z), (D, X)\}$.

9. The flow augmentation algorithm shows that the given matching is a maximum. (The vertices A, B, and D are joined only to X and Z.)

11. The flow augmentation algorithm gives the maximum matching $\{(A, 1), (C, 3), (D, 2)\}$.

13. The flow augmentation algorithm gives the maximum matching

$$\{(a, A), (b, C), (c, B), (d, D)\}.$$

15. A matching with four edges exists. One possible team is: Andrew and Greta, Bob and Hannah, Dan and Flo, Ed and Iris.

17. A matching with five edges exists. One possible assignment is: Craig files, Dianne distributes paychecks, Gale collates, Marilyn types, and Sharon helps students.

19. Create a bipartite graph $\mathcal{G}$ with vertices U_i $(1 \leq i \leq n)$ that correspond to the sets S_i and vertices V_j that correspond to the elements in $S_1 \cup S_2 \cup \cdots \cup S_n$. Join U_i and V_j if and only if the element corresponding to V_j belongs to set S_i. Apply the flow augmentation algorithm to the network $\mathcal{N}$ associated with $\mathcal{G}$. Then the sequence $S_1 \cup S_2 \cup \cdots \cup S_n$ has a system of distinct representatives if and only if the value of a maximal flow in $\mathcal{N}$ is n.

21. No system of distinct representatives exists because sets 1, 3, 5, and 6 collectively contain only three elements.

23. No acceptable assignment exists because there are only three professors who are capable of teaching British History, Latin American History, Oliver Cromwell, and 20th Century History.

25. Let $\mathcal{A} = \varnothing \subseteq \mathcal{V}_1$. Then $\mathcal{A}^* = \varnothing$, and so

$$|\mathcal{A}| - |\mathcal{A}^*| = 0 - 0 = 0.$$

Because d is the maximum of $|\mathcal{A}| - |\mathcal{A}^*|$ over all subsets $\mathcal{A}$ of $\mathcal{V}_1$, we have $d \geq 0$.

27. Clearly $s \in \mathcal{S}$, $t \in \mathcal{T}$, and every vertex in $\mathcal{G}$ lies in exactly one of $\mathcal{S}$ and $\mathcal{T}$. So $\mathcal{S}, \mathcal{T}$ is a cut. The only vertices in $\mathcal{G}$ that are adjacent to vertices in $\mathcal{A}$ belong to $\mathcal{A}^*$, and so there are no arcs from a vertex in $\mathcal{A}$ to a vertex in $\mathcal{V}_2 - \mathcal{A}^*$. Moreover, in the construction of $\mathcal{N}$, there are no arcs from a vertex in $\mathcal{V}_1$ to another vertex in $\mathcal{V}_1$, from a vertex in $\mathcal{V}_2$ to another vertex in $\mathcal{V}_2$, or from a vertex in $\mathcal{V}_2$ to a vertex in $\mathcal{V}_1$. Hence the only arcs from $\mathcal{S}$ to $\mathcal{T}$ are those from s to vertices in $\mathcal{V}_1 - \mathcal{A}$ and those from vertices in $\mathcal{A}^*$ to t. The number of such arcs is

$$|\mathcal{V}_1 - \mathcal{A}| + |\mathcal{A}^*| = |\mathcal{V}_1| - |\mathcal{A}| + |\mathcal{A}^*|$$
$$= |\mathcal{V}_1| - (|\mathcal{A}| - |\mathcal{A}^*|)$$
$$= |\mathcal{V}_1| - d.$$

Since each arc has capacity 1, the capacity of the cut $\mathcal{S}, \mathcal{T}$ is $|\mathcal{V}_1| - d$.

29. If $f(X, Y) = 1$, then we must have $f(s, X) = 1$ because (s, X) is the only arc leading into X. To prove that X is unlabeled or Y is labeled, it suffices to show that whenever Y is unlabeled, then X is unlabeled. Suppose then that Y is unlabeled. Then X could not have been labeled by scanning from Y. Moreover, since there is no flow along any arc (X, Y') for $Y' \in \mathcal{V}_2 - \{Y\}$, X could not have been labeled by scanning from any vertex in $\mathcal{V}_2$. Thus the only way in which X could have been labeled is by scanning from s, which is impossible because the flow along arc (s, X) equals its capacity. Hence X is unlabeled or Y is labeled.

31. Suppose that there is an edge in $\mathcal{G}$ joining a labeled vertex $X \in \mathcal{V}_1$ to an unlabeled vertex $Y \in \mathcal{V}_2$. Then in the network associated with $\mathcal{G}$, there is an arc from X to Y. Since X is labeled and Y is unlabeled, the flow along arc (X, Y) must equal its capacity, which is 1. But this is a contradiction of Exercise 29. So there can be no edge in $\mathcal{G}$ joining a labeled vertex $X \in \mathcal{V}_1$ to an unlabeled vertex $Y \in \mathcal{V}_2$. Thus every edge in $\mathcal{G}$ must be incident with an unlabeled vertex in $\mathcal{V}_1$ or a labeled vertex in $\mathcal{V}_2$. It follows that the set containing the unlabeled vertices in $\mathcal{V}_1$ and the labeled vertices in $\mathcal{V}_2$ is a covering of $\mathcal{G}$ in the sense of Section 6.2. By Exercise 30, the number of vertices in this covering equals the number of edges in a maximum matching of $\mathcal{G}$, and so this covering must be a minimum covering by Theorem 6.2.

SUPPLEMENTARY EXERCISES

1. In the first iteration of the flow augmentation algorithm, 4 units will be sent along the directed path A, C, F. In the second iteration, 5 units will be sent along A, B, D, F. In the third iteration, 2 units will be sent along A, B, E, F. In the fourth iteration, 4 units will be sent along A, C, D, E, F. The resulting maximal flow (shown below) has a value of 15; a minimal cut is $\{A\}, \{B, C, D, E, F\}$.

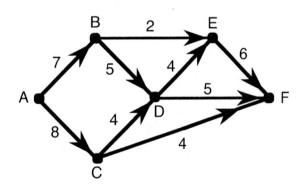

3. In the first iteration of the flow augmentation algorithm, 6 units will be sent along the directed path A, D, G. In the second iteration, 4 units will be sent along A, B, E, G. In the third iteration, 2 units will be sent along A, C, F, G. In the fourth iteration, 3 units will be sent along A, B, D, E, G. In the fifth iteration, 1 unit will be sent along A, B, D, F, G. In the sixth iteration, 3 units will be sent along the path A, C, D, F, G. The maximal flow (shown on the next page) has a value of 19, and a minimal cut is

$$\{A, B, C\}, \{D, E, F, G\}.$$

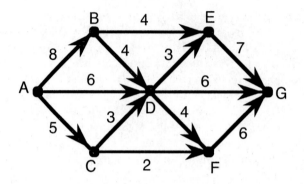

5. In the first iteration of the flow augmentation algorithm, 3 units will be sent along the directed path A, B, E, H. In the second iteration, 3 units will be sent along A, B, F, H. In the third iteration, 4 units will be sent along A, C, E, H. In the fourth iteration, 1 unit will be sent along A, D, E, H. In the fifth iteration, 3 units will be sent along A, D, G, H. In the sixth iteration, 2 units will be sent along the path A, C, B, F, H. In the seventh iteration, 3 units will be sent along A, D, E, G, H. The resulting maximal flow has a value of 19, and a minimal cut is $\{A, B, C, D, F\}$, $\{E, G, H\}$. The maximal flow is shown below.

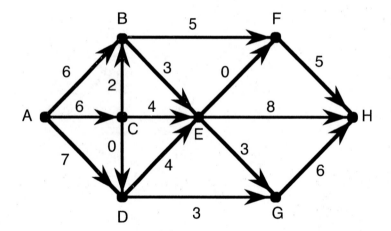

7. In the first iteration of the flow augmentation algorithm, 3 units will be sent along the directed path A, B, E, H, K. In the second iteration, 6 units will be sent along A, B, F, H, K. In the third iteration, 1 unit will be sent along A, B, F, I, K. In the fourth iteration, 4 units will be sent along A, C, F, I, K. In the fifth iteration, 4 units will be sent along A, C, F, J, K. In the sixth iteration, 1 unit will be sent along A, C, G, J, K. In the seventh iteration, 2 units will be sent along A, D, G, J, K. In the eighth iteration, 3 units will be sent along A, D, G, J, I, K. In the ninth iteration, 1 unit will be sent along A, D, C, F, H, I, K. In the tenth iteration, 2 units will be

sent along A, D, C, G, J, I, K. In the eleventh iteration, 1 unit will be sent along $A, D, C, G, J, F, H, I, K$. The resulting maximal flow has a value of 28, and a minimal cut is $\{A, D\}$, $\{B, C, E, F, G, H, I, J, K\}$. The maximal flow is shown below.

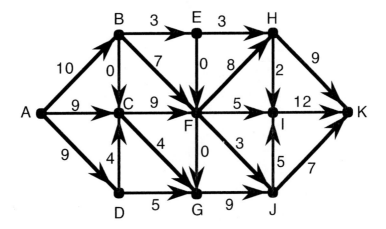

9. By taking $\mathcal{S}_0 = \{v_1, v_2, \ldots, v_r\}$ and imitating the proof of Theorem 7.1, we see that $O_1 + O_2 + \cdots + O_r = I_n$. Thus the value of the flow equals the total flow out of all the vertices in $\mathcal{S}_0$.

11. Applying the flow augmentation algorithm to the transportation network described in Exercise 10 produces the following results. In the first iteration, 3 units will be sent along the path u, A, D, G. In the second iteration, 4 units will be sent along the path u, A, E, G. In the third iteration, 1 unit will be sent along the path u, B, E, G. In the fourth iteration, 4 units will be sent along the path u, B, F, G. In the fifth iteration, 1 unit will be sent along the path u, C, D, G. In the sixth iteration, 4 units will be sent along the path u, C, F, G. The resulting maximal flow is shown below.

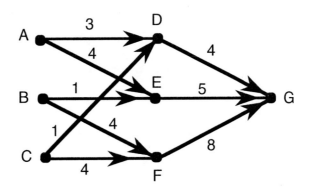

13. Modify the definition of a multisource transportation network to allow a nonempty finite set $\mathcal{T}_0$ of vertices with outdegree 0. A **flow** f must satisfy $0 \le f(e) \le c(e)$ for each arc e, and the total flow into each vertex not in $\mathcal{S}_0 \cup \mathcal{T}_0$ must equal the total flow out of that vertex. The **value** of a flow is the total flow into elements of $\mathcal{T}_0$. A **maximal flow** in such a network is a flow having the largest possible value.

Theorem For a network and flow as defined above, the total flow out of $\mathcal{S}_0$ equals the total flow into $\mathcal{T}_0$.

Proof Introduce a new source s and arcs of infinite capacity from s to each vertex in $\mathcal{S}_0$ and a new sink t and arcs of infinite capacity from each vertex in $\mathcal{T}_0$ to t. By Theorem 7.1, the total flow out of s must equal the total flow into t. Thus the total flow out of $\mathcal{S}_0$ equals the total flow into $\mathcal{T}_0$. $\square$

15. The network $\mathcal{N}^*$ described in Exercise 14 is shown below.

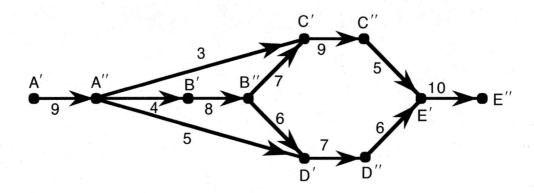

17. The network $\mathcal{N}^*$ described in Exercise 14 is shown below.

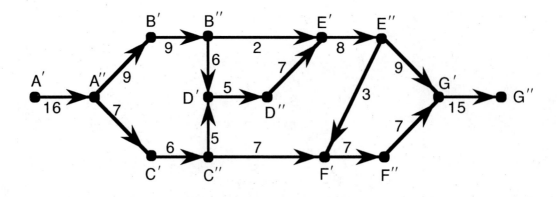

19. When the flow augmentation algorithm is applied to the network in Exercise 16, 3 units are sent along the directed path $A', A'', B', B'', D', D'', F', F''$ in the first iteration; 3 units are sent along $A', A'', C', C'', E', E'', F', F''$ in the second iteration; and 1 unit is sent along $A', A'', C', C'', B', B'', D', D'', F', F''$ in the third iteration. The resulting flow in the original network with vertex capacities is shown below.

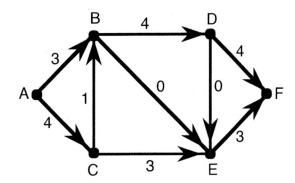

21. Define an $m \times n$ matrix A by $A_{i,j} = f(x_i, y_j)$ for $1 \le i \le m$ and $1 \le j \le n$. In this transportation network, the only arcs coming into x_i begin at s, and the only arcs going out of x_i end at some y_j. Hence the number of 1s in row i of A is

$$A_{i1} + A_{i2} + \cdots + A_{in} = f(x_i, y_1) + f(x_i, y_2) + \cdots + f(x_i, y_n) = f(s, x_i)$$

because the total flow out of x_i equals the total flow into x_i. Likewise the number of 1s in column j of A is

$$A_{1j} + A_{2j} + \cdots + A_{mj} = f(x_1, y_j) + f(x_2, y_j) + \cdots + f(x_m, y_j) = f(y_j, t).$$

23. Let f be a maximal flow in the network, and assume that $u_1 + u_2 + \cdots + u_m$ is the value of f. Define $A_{i,j} = f(x_i, y_j)$ for $1 \le i \le m$ and $1 \le j \le n$. As in Exercise 21, the number of 1s in row i of A is

$$A_{i1} + A_{i2} + \cdots + A_{in} = f(x_i, y_1) + f(x_i, y_2) + \cdots + f(x_i, y_n) = f(s, x_i) = u_i,$$

and the number of 1s in column j of A is

$$A_{1j} + A_{2j} + \cdots + A_{mj} = f(x_1, y_j) + f(x_2, y_j) + \cdots + f(x_m, y_j) = f(y_j, t) = v_j.$$

25. In the network below, each arc is directed from left to right, and every arc from some x_i to a y_j has capacity 1.

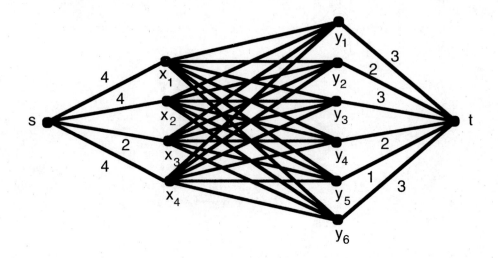

Apply the flow augmentation algorithm to this network until a maximal flow f is obtained. Then define $A_{i,j} = f(x_i, y_j)$ as in Exercise 23. The resulting 4×6 matrix is

$$A = \begin{bmatrix} 1 & 1 & 1 & 0 & 0 & 1 \\ 1 & 1 & 1 & 0 & 0 & 1 \\ 1 & 0 & 0 & 1 & 0 & 0 \\ 0 & 0 & 1 & 1 & 1 & 1 \end{bmatrix}.$$

Chapter 8

Counting Techniques

8.1 PASCAL'S TRIANGLE AND THE BINOMIAL THEOREM

1. $C(5,3) = \dfrac{5!}{3!\,(5-3)!} = \dfrac{5\cdot4\cdot3!}{3!\,2!} = \dfrac{5\cdot4}{2\cdot1} = \dfrac{20}{2} = 10$

3. $C(8,5) = \dfrac{8!}{5!\,(8-5)!} = \dfrac{8\cdot7\cdot6\cdot5!}{5!\,3!} = \dfrac{8\cdot7\cdot6}{3\cdot2\cdot1} = 56$

5. By the binomial theorem, this coefficient is

$$C(4,2) = \frac{4!}{2!\,(4-2)!} = \frac{4\cdot3\cdot2!}{2!\,2!} = \frac{4\cdot3}{2\cdot1} = 6.$$

7. By the binomial theorem, this coefficient is $C(12,9)(-1)^9 = -C(12,9) = -220$.

9. By the binomial theorem, this coefficient is $C(10,4)(2)^4 = 210(16) = 3360$.

11. By the binomial theorem, this coefficient is $C(10,7)(-3)^7 = 120(-2187) = -262{,}440$.

13. The $n = 6$ row of Pascal's triangle can be computed as in Example 7.3. It is:

$$1, \quad 1+5 = 6, \quad 5+10 = 15, \quad 10+10 = 20, \quad 10+5 = 15, \quad 5+1 = 6, \quad 1.$$

15. The binomial theorem gives

$$(x+y)^6 = x^6 + 6x^5y + 15x^4y^2 + 20x^3y^3 + 15x^2y^4 + 6xy^5 + y^6.$$

17. The binomial theorem gives

$$(3x-y)^4 = (3x)^4 + 4(3x)^3(-y) + 6(3x)^2(-y)^2 + 4(3x)(-y)^3 + (-y)^4$$
$$= 81x^4 - 108x^3y + 54x^2y^2 - 12xy^3 + y^4.$$

19. The number of 4-element subsets that can be formed from a set of 7 elements is $C(7,4) = 35$.

21. The number of 5-element subsets that can be formed from a set of 10 elements is $C(10,5) = 252$.

23. The number of 4-element subsets that can be formed from the set $\{2, 4, 6, 8, 10, 12\}$ is $C(6,4) = 15$.

25. Applying the binomial theorem to $(1+1)^n$ gives

$$2^n = (1+1)^n$$
$$= C(n,0)(1^n) + C(n,1)(1^{n-1})(1) + C(n,2)(1^{n-2})(1^2) + \cdots + C(n,n)(1^n)$$
$$= C(n,0) + C(n,1) + C(n,2) + \cdots + C(n,n).$$

27. Applying the binomial theorem to $(1-1)^n$ gives

$$0 = 0^n$$
$$= (1-1)^n$$
$$= C(n,0)(1^n) + C(n,1)(1^{n-1})(-1) + C(n,2)(1^{n-2})(-1)^2 + \cdots + C(n,n)(-1)^n$$
$$= C(n,0) - C(n,1) + C(n,2) - \cdots + (-1)^n C(n,n).$$

29. We have

$$r \cdot C(n,r) = r\frac{n!}{r!\,(n-r)!} = \frac{n!}{(r-1)!\,(n-r)!} = \frac{n(n-1)!}{(r-1)!\,(n-r)!} = n \cdot C(n-1, r-1).$$

31. Let k be a positive integer. The proof will be by induction on r. If $r = 0$, then:

$$C(k,0) + \cdots + C(k+r,r) = C(k,0) = 1 = C(k+1,0) = C(k+r+1, r).$$

Hence the equation holds for $r = 0$. Assume that the equation holds for $r = n$, where n is a nonnegative integer. Then, by the induction hypothesis and Theorem 8.1, we have:

$$C(k,0) + C(k+1,1) + \cdots + C(k+n,n) + C(k+n+1, n+1)$$
$$= C(k+n+1, n) + C(k+n+1, n+1)$$
$$= C(k+n+2, n+1).$$

Hence the equation holds for $r = n + 1$. Thus, by the principle of mathematical induction, the equation holds for all nonnegative integers r.

33. Since $C(2n+1, r) = C(2n+1, 2n+1-r)$, it follows from Exercise 25 that

$$2[C(2n+1, 0) + C(2n+1, 1) + \cdots + C(2n+1, n)]$$
$$= C(2n+1, 0) + \cdots + C(2n+1, n) +$$
$$C(2n+1, n+1) + \cdots + C(2n+1, 2n+1)$$
$$= 2^{2n+1}.$$

Thus

$$C(2n+1, 0) + C(2n+1, 1) + \cdots + C(2n+1, n) = \frac{2^{2n+1}}{2} = 2^{2n}.$$

35. By (8.1), we have

$$C(n, k+1) = \frac{n!}{(k+1)!\,(n-k-1)!} = \frac{n-k}{k+1} \cdot \frac{n!}{k!\,(n-k)!} = \frac{n-k}{k+1} \cdot C(n, k).$$

If $k < n/2$, then

$$n - k > n - \frac{n}{2} = \frac{n}{2} > k.$$

Hence

$$\frac{n-k}{k+1} \geq 1.$$

Thus

$$C(n, k+1) = \frac{n-k}{k+1} \cdot C(n, k) \geq C(n, k).$$

37. Let n be a positive integer. For any product $k(k+1)\cdots(k+n-1)$ of n positive integers, we have

$$\frac{k(k+1)\cdots(k+n-1)}{n!} = \frac{(k+n-1)!}{(k-1)!\,n!} = C(k+n-1, n),$$

which is an integer.

8.2   THREE FUNDAMENTAL PRINCIPLES

1. Since there are twelve months in a year, it follows from the pigeonhole principle that, in any set of 13 people, at least two are born in the same month.

3. Since there are socks of four different colors, it follows from the pigeonhole principle that, in any set of 5 socks, at least two must have the same color.

5. It follows from the pigeonhole principle that there must be at least 14 chairs at the table having the most seats. (Otherwise, there can be no more than $8 \cdot 13 = 104$ seats.)

7. Since there are four different kinds of books, the pigeonhole principle implies that a minimum of $4 \cdot (12 - 1) + 1 = 45$ books must be chosen in order to assure that there are at least 12 books on the same subject.

9. Since there are 5 different options, each of which is to be included or excluded, the multiplication principle shows that the number of different sets of options is $2^5 = 32$.

11. Every subset of a set containing n objects can be formed by choosing, for each object in the set, to include the object in the subset or to exclude the object from the subset. The number of different subsets that can be formed is therefore the number of different ways to make these choices. The multiplication principle shows that this number is $2 \cdot 2 \cdots 2 = 2^n$.

13. The number of different routings equals the product of the number of flights from Kansas City to Chicago and the number of flights from Chicago to Boston, which is $8 \cdot 21 = 168$.

15. There are 8 choices for the first digit (2 through 9), 2 choices for the second digit (0 or 1), and 10 choices for the third digit (0 through 9). Hence the number of possible area codes is $8 \cdot 2 \cdot 10 = 160$.

17. (a) If anyone may sit in any chair, there are $6 \cdot 5 \cdot 4 \cdot 3 \cdot 2 \cdot 1 = 720$ different seating arrangements.

(b) If men must occupy the first and last chairs, there are 3 choices of person to fill the first chair. After the first chair is filled, there are 2 persons left who may be chosen to fill the last chair. We are left with 4 persons who may be chosen to fill the second chair, 3 persons who may be chosen to fill the third chair, 2 persons who may be chosen to fill the fourth chair, and 1 person who may be chosen to fill the fifth chair. Thus the number of different seating arrangements is $3 \cdot 4 \cdot 3 \cdot 2 \cdot 1 \cdot 2 = 144$.

(c) If men must occupy the first three chairs, there are $3 \cdot 2 \cdot 1 \cdot 3 \cdot 2 \cdot 1 = 36$ different seating arrangements.

(d) If everyone must be seated beside his or her spouse, there are $6 \cdot 1 \cdot 4 \cdot 1 \cdot 2 \cdot 1 = 48$ different seating arrangements.

19. The number of different identifiers containing exactly two characters is $52 \cdot 62 - 5 = 3219$.

21. By the addition principle, the number of different ways to select an item on sale is $30 + 40 = 70$.

23. As in Example 8.14, we see that the number of 8-bit strings beginning with 1001 is $2^4 = 16$ and the number beginning with 010 is $2^5 = 32$. Since no string can begin with both 1001 and 010, the number of 8-bit strings beginning with 1001 or 010 is $16 + 32 = 48$ by the addition principle.

25. The number of 3-letter combinations is $2 \cdot 26 \cdot 26 = 1352$, and the number of 4-letter combinations is $2 \cdot 26 \cdot 26 \cdot 26 = 35{,}152$. Hence the number of 3-letter or 4-letter combinations is $1{,}352 + 35{,}152 = 36{,}504$.

27. **(a)** There are $10 \cdot 9 \cdot 8 = 720$ possible appointments.

(b) There are $4 \cdot 6 \cdot 5$ appointments in which a man is sent to the first city. Likewise there are $6 \cdot 4 \cdot 5$ appointments in which a man is sent to the second city and $6 \cdot 5 \cdot 4$ appointments in which a man is sent to the third city. Hence the number of possible appointments is $4 \cdot 6 \cdot 5 + 6 \cdot 4 \cdot 5 + 6 \cdot 5 \cdot 4 = 360$.

(c) If at least two men must be appointed, we must distinguish between the cases where exactly two men and exactly three men are appointed. The case in which exactly two men are appointed is handled as in (b). In all, there are

$$(4 \cdot 3 \cdot 6 + 4 \cdot 6 \cdot 3 + 6 \cdot 4 \cdot 3) + 4 \cdot 3 \cdot 2 = 240$$

possible appointments.

(d) If at least one person of each sex must be appointed, we must distinguish between the cases where exactly two men and exactly two women are appointed. As in (b), there are

$$(6 \cdot 5 \cdot 4 + 6 \cdot 4 \cdot 5 + 4 \cdot 6 \cdot 5) + (4 \cdot 3 \cdot 6 + 4 \cdot 6 \cdot 3 + 6 \cdot 4 \cdot 3) = 576$$

possible appointments.

29. This question is similar to Example 8.14(b). The number of strings that begin with 11 and do not end with 00 is $2^4 \cdot 3$. Likewise the number of strings that end with 00 and do not begin with 11 is $3 \cdot 2^4$, and the number of strings that begin with 11 and end with 00 is 2^4. Hence the number of strings that begin with 11 or end with 00 is $2^4 \cdot 3 + 3 \cdot 2^4 + 2^4 = 112$.

31. **(a)** There are $6^4 = 1296$ numbers if repetition is allowed.

(b) There are $6 \cdot 5 \cdot 4 \cdot 3 = 360$ numbers if repetition is not allowed.

(c) Of the numbers in (b), $1 \cdot 5 \cdot 4 \cdot 3 = 60$ begin with 3.

(d) We can choose a position for the 2 in 4 ways. The remaining digits can be chosen in $5 \cdot 4 \cdot 3 = 60$ ways. Thus there are $4 \cdot 60 = 240$ numbers in (b) that contain 2.

33. There are 5 vowels and 21 consonants in the English alphabet. In a circular array, each of the 26 letters begins a sequence of 5 letters, and each vowel occurs in 5 different sequences (once in each possible position). Hence at most $5 \cdot 5 = 25$ sequences contain a vowel. It follows that some sequence contains no vowels.

35. There are 8 parity classes for integer points in Euclidean space, namely:

(even, even, even);   (even, even, odd);   (even, odd, even);   (even, odd, odd);

(odd, even, even);   (odd, even, odd);   (odd, odd, even);   and   (odd, odd, odd).

159

Thus in any set of 9 integer points there must be two in the same parity class. But the midpoint of any two points in the same parity class will have only integer coordinates.

8.3 PERMUTATIONS AND COMBINATIONS

1. $C(6,3) = \dfrac{6!}{3!\,(6-3)!} = \dfrac{6\cdot 5\cdot 4\cdot 3!}{3!\,3!} = \dfrac{6\cdot 5\cdot 4}{3\cdot 2\cdot 1} = 5\cdot 4 = 20$

3. $C(5,2) = \dfrac{5!}{2!\,(5-2)!} = \dfrac{5\cdot 4\cdot 3!}{2!\,3!} = \dfrac{5\cdot 4}{2\cdot 1} = \dfrac{20}{2} = 10$

5. $P(4,2) = \dfrac{4!}{(4-2)!} = \dfrac{4\cdot 3\cdot 2!}{2!} = 4\cdot 3 = 12$

7. $P(9,5) = \dfrac{9!}{(9-5)!} = \dfrac{9\cdot 8\cdot 7\cdot 6\cdot 5\cdot 4!}{4!} = 9\cdot 8\cdot 7\cdot 6\cdot 5 = 15{,}120$

9. $P(10,4) = \dfrac{10!}{(10-4)!} = \dfrac{10\cdot 9\cdot 8\cdot 7\cdot 6!}{6!} = 10\cdot 9\cdot 8\cdot 7 = 5040$

11. $P(n,1) = \dfrac{n!}{(n-1)!} = \dfrac{n(n-1)!}{(n-1)!} = n$

13. $P(4,4) = 4! = 24$

15. $P(6,4) = 360$

17. $C(13,3) = 286$

19. $C(10,6) = 210$

21. $P(5,5) = 5! = 120$

23. $C(9,3) = 84$

25. $C(6,3)\cdot C(5,2) = 20(10) = 200$

27. $P(9,3)\cdot C(7,2) = 504(21) = 10{,}584$

29. **(a)** $C(12,4) = 495$
(b) $C(5,4) = 5$
(c) $C(3,1)\cdot C(4,1)\cdot C(2,1)\cdot C(3,1) = 3\cdot 4\cdot 2\cdot 3 = 72$
(d) $C(3,1)\cdot C(4,2)\cdot C(3,1) = 3(6)(3) = 54$

31. Two elements can be chosen from among $2n$ elements $a_1, a_2, \ldots, a_{2n}$ in any of the following three ways.

(i) Choose two elements from among $a_1, a_2, \ldots, a_n$.

(ii) Choose two elements from among $a_{n+1}, a_{n+2}, \ldots, a_{2n}$.

(iii) Choose one element from among $a_1, a_2, \ldots, a_n$ and one element from among $a_{n+1}, a_{n+2}, \ldots, a_{2n}$.

Thus the total number of ways to choose two elements from among $2n$ elements is

$$C(n, 2) + C(n, 2) + n \cdot n = 2C(n, 2) + n^2.$$

33. Note that $C(n, m)$ counts the number of ways to choose a set of m finalists from a set of n contestants, and $C(m, k)$ counts the number of ways to choose a set of k winners from the set of m finalists. Thus the left side of the equation counts the number of ways to choose a set of m finalists from a set of n contestants and then choose a set of k winners from among the finalists. On the other hand, the right side of the equation counts the number of ways to choose a set of k winners from among n contestants and then choose a set of $m - k$ nonwinning finalists from the remaining $n - k$ contestants. Clearly these numbers are equal.

35. In order to prove this equality by a combinatorial argument, let us suppose that we choose 2 elements from among $a_1, a_2, \ldots, a_{n+1}$. The number of such choices is $C(n + 1, 2)$. Now we consider another method for choosing two of these elements, say a_i and a_j, where $i < j$. Clearly $2 \leq j \leq n + 1$. If $j = 2$, then there is $C(1, 1)$ way to choose a_i. If $j = 3$, then there are $C(2, 1)$ ways to choose a_i. More generally, if $j = k$, there are $C(k - 1, 1)$ ways to choose a_i. Since we can choose 2 elements from among $a_1, a_2, \ldots, a_{n+1}$ by first choosing a_j with $2 \leq j \leq n+1$ and then choosing a_i such that $1 \leq i < j$, it follows that

$$C(n + 1, 2) = C(1, 1) + C(2, 1) + \ldots + C(n, 1).$$

8.4   ARRANGEMENTS AND SELECTIONS WITH REPETITIONS

1. Since "redbird" contains two r's, one e, two d's, one b, and one i, there are

$$\frac{7!}{2! \, 1! \, 2! \, 1! \, 1!} = 1260$$

distinct rearrangements of its letters by Theorem 8.7.

3. Since 5,363,565 contains three 5's, two 3's, and two 6's, by Theorem 8.7 there are

$$\frac{7!}{3! \, 2! \, 1!} = 210$$

distinct rearrangements of its digits.

161

5. Because each fruit basket can be formed by choosing 8 pieces of fruit with repetition from among three types (apples, oranges, and pears), there are

$$C(8 + 3 - 1, 8) = C(10, 8) = 45$$

different fruit baskets possible by Theorem 8.8.

7. Each assortment of one dozen donuts can be formed by choosing 12 donuts with repetition from among the 5 available types. The number of possible assortments is therefore

$$C(12 + 5 - 1, 12) = C(16, 12) = 1820$$

by Theorem 8.8.

9. By Theorem 8.7, the number of ways to distribute 16 crayons (no two of the same color) to four children so that each child receives 4 crayons is

$$C(16, 4) \cdot C(12, 4) \cdot C(8, 4) \cdot C(4, 4) = \frac{16!}{4!\,4!\,4!\,4!} = 63{,}063{,}000.$$

11. Each subdivision of the telephone calls between the chairperson and the secretary can be found by ordering the names of the persons to be called and having the chairperson call the first three and the secretary call the last four. Hence, by Theorem 8.7, the number of possible subdivisions is

$$\frac{7!}{3!\,4!} = 35.$$

13. The number of possible distributions is $C(8 + 4 - 1, 8) = C(11, 8) = 165$.

15. If each student must receive at least one stick of chalk, give one stick to each student. By Theorem 8.8, the remaining three sticks can be distributed in $C(3 + 3 - 1, 3) = C(5, 3) = 10$ different ways.

17. The number of possible programs equals the number of ways to order 1 Baroque piece, 3 classical pieces, and 3 romantic pieces. By Theorem 8.7, this number is

$$\frac{7!}{1!\,3!\,3!} = 140.$$

19. By Theorem 8.8, the number of ways to distribute the mathematics books is

$$C(8 + 6 - 1, 8) = C(13, 8),$$

and the number of ways to distribute the computer science books is

$$C(10 + 6 - 1, 10) = C(15, 10).$$

Hence, by the multiplication principle, the number of ways that both types of books can be distributed is

$$C(13, 8) \cdot C(15, 10) = 1287 \cdot 3003 = 3{,}864{,}861.$$

21. By Theorem 8.7, the number of different assignments is

$$\frac{10!}{3!\,4!\,3!} = 4200.$$

23. Arrange the ten 1s in a row and insert the six 0s among the 1s so that there are no two consecutive 0s. This amounts to choosing positions for the six 0s from among eleven possible locations (before the first 1, between consecutive 1s, and after the last 1). The number of ways to choose locations for the 0s is $C(11,6) = 462$.

25. There are two cases to consider. If some child receives two teddy bears, then, by Theorem 8.7, the remaining 7 distinct dolls can be distributed in

$$C(7,1) \cdot C(6,3) \cdot C(3,3) = \frac{7!}{1!\,3!\,3!} = 140$$

different ways. Since there are $C(3,1) = 3$ possible choices for the child who receives the teddy bears, the number of possible distributions in this case is $3 \cdot 140 = 420$.

If two children each receive one teddy bear, then, by Theorem 8.7, the remaining 7 distinct dolls can be distributed in

$$C(7,2) \cdot C(5,2) \cdot C(3,3) = \frac{7!}{2!\,2!\,3!} = 210$$

different ways. Since there are $C(3,2) = 3$ possible choices for the children who receive the teddy bears, the number of possible distributions in this case is $3 \cdot 210 = 630$. Thus there are $420 + 630 = 1050$ possible distributions in all.

27. A number less than 10,000 has no more than 4 digits. Furthermore if the digits of the number sum to 8, then they can be chosen by distributing eight identical 1s into four distinct boxes marked *ones digit*, *tens digit*, *hundreds digit*, and *thousands digit*. By Theorem 8.8, the number of such distributions is $C(8+4-1,8) = C(11,8) = 165$. So there are 165 numbers less than 10,000 in which the digits sum to 8.

29. Consider a positive integer less than 1,000,000 for which the sum of the digits equals 12. Let the digits of this number be x_1, x_2, x_3, x_4, x_5, and x_6. Then

$$x_1 + x_2 + x_3 + x_4 + x_5 + x_6 = 12$$

and $0 \leq x_i \leq 9$. Therefore a positive integer less than 1,000,000 for which the sum of the digits equals 12 is a nonnegative integer solution of the equation

$$x_1 + x_2 + x_3 + x_4 + x_5 + x_6 = 12$$

with $0 \leq x_i \leq 9$. By Theorem 8.8, there are $C(12+6-1,12)$ nonnegative integer solutions to this equation. Note that it is impossible for more than one x_i to exceed 9 if the sum of the x_i equals 12. Furthermore, the number of such solutions in which a particular x_i is 10 or more is $C(2 + 6 - 1, 2)$. Hence the number of solutions in which some x_i is 10 or more is $6 \cdot C(2 + 6 - 1, 2)$. Thus the number of nonnegative integer solutions of the equation $x_1 + x_2 + x_3 + x_4 + x_5 + x_6 = 12$ such that $0 \leq x_i \leq 9$ is

$$C(12 + 6 - 1, 12) - 6 \cdot C(2 + 6 - 1, 2) = 6188 - 6 \cdot 21 = 6062.$$

31. Each line that is printed consists of three integers I, J, K, where $10 \geq I \geq J \geq K \geq 1$. Moreover, every such sequence of three integers is printed. Thus the number of times that the PRINT statement is executed equals the number of sequences of this form. But each of these sequences is simply a selection of three integers with repetition chosen from among $\{1, 2, \ldots, 10\}$ and written in nonincreasing order. By Theorem 8.8, the number of such sequences is $C(3 + 10 - 1, 3) = C(12, 3) = 220$.

33. The number of duplicate cards that occur in the hand must be 0, 1, 2, 3, 4, 5, or 6. If there are no duplicate cards in the hand, then the number of different ways in which the cards can be chosen is $C(24, 12)$. If there is one duplicate card in the hand, then there are $C(24, 1)$ ways to choose the duplicated card and $C(23, 10)$ ways to choose the remaining ten cards in the hand. Similar reasoning shows that the number of different pinochle hands is

$$C(24, 12) + C(24, 1) \cdot C(23, 10) + C(24, 2) \cdot C(22, 8) + C(24, 3) \cdot C(21, 6)$$
$$+ C(24, 4) \cdot C(20, 4) + C(24, 5) \cdot C(19, 2) + C(24, 6)$$

$$= 2{,}704{,}156 + 24 \cdot 1{,}144{,}066 + 276 \cdot 319{,}770 + 2024 \cdot 54{,}264$$
$$+ 10{,}626 \cdot 4845 + 42{,}504 \cdot 171 + 134{,}596$$

$$= 287{,}134{,}346.$$

35. We have

$$C(n, n_1) \cdot C(n - n_1, n_2) \cdot C(n - n_1 - n_2, n_3) \cdots C(n - n_1 - n_2 - \cdots - n_{k-1}, n_k)$$

$$= \frac{n!}{n_1! \, (n - n_1)!} \cdot \frac{(n - n_1)!}{n_2! \, (n - n_1 - n_2)!} \cdots \frac{(n - n_1 - n_2 - \cdots - n_{k-1})!}{n_k! \, (n - n_1 - n_2 - \cdots - n_{k-1})!}$$

$$= \frac{n!}{n_1! \, n_2! \cdots n_k!}.$$

8.5 PROBABILITY

1. Since there are 6 numbers that can be rolled and 5 of them are greater than 1, the probability of rolling a number greater than 1 is 5/6.

3. Because there are $2^5 = 32$ possible sequences of heads and tails and in only one of them does the coin land heads each time, the probability that the coin lands heads on 5 consecutive tosses is $1/32$.

5. Since there are $6^2 = 36$ outcomes when a pair of dice are rolled and only two ways to obtain a sum of 11 (namely, $5 + 6$ and $6 + 5$), the probability of rolling a sum of 11 on a pair of dice is
$$\frac{2}{36} = \frac{1}{18}.$$

7. The probability that 3 of the five coins land tails is
$$\frac{C(5,3)}{2^5} = \frac{10}{32} = \frac{5}{16}.$$

9. The probability that 3 women are chosen is
$$\frac{C(6,3)}{C(11,3)} = \frac{20}{165} = \frac{4}{33}.$$

11. The probability of guessing the first three horses in order is
$$\frac{1}{P(7,3)} = \frac{1}{210}.$$

13. The probability that a randomly chosen four-digit number contains no repeated digits is
$$\frac{9 \cdot 9 \cdot 8 \cdot 7}{9 \cdot 10 \cdot 10 \cdot 10} = \frac{63}{125}.$$

15. If the letters of the word "sassafras" are randomly permuted, the probability that the four s's are adjacent and the three a's are adjacent is
$$\frac{\left(\dfrac{4!}{1!\,1!\,1!\,1!}\right)}{\left(\dfrac{9!}{4!\,3!\,1!\,1!}\right)} = \frac{1}{105}.$$

17. The probability of selecting five employees' files in order of increasing salary is
$$\frac{1}{P(5,5)} = \frac{1}{120}.$$

19. The number of subsets of $\{1,2,3,4,5,6\}$ that contain both 3 and 5 equals the number of subsets of $\{1,2,4,6\}$, which is 2^4. Since there are 2^6 subsets of $\{1,2,3,4,5,6\}$, the probability that a randomly chosen subset of $\{1,2,3,4,5,6\}$ contains both 3 and 5 is
$$\frac{2^4}{2^6} = \frac{16}{64} = \frac{1}{4}.$$

21. Regard the sticks of gum as distinct. If each child receives at least four sticks of gum, then some child receives five sticks and the other two receive four sticks each. The number of ways to distribute 13 distinct sticks of gum to 3 distinct children is 3^{13}, and the number of such distributions in which a particular child receives five sticks and the other two receive four sticks is

$$C(13,5) \cdot C(8,4) \cdot C(4,4).$$

Since there are $C(3,1)$ ways to choose which child receives five sticks of gum, the probability that each child receives at least four sticks of gum is

$$\frac{C(3,1) \cdot C(13,5) \cdot C(8,4) \cdot C(4,4)}{3^{13}} = \frac{3 \cdot 1287 \cdot 70 \cdot 1}{3^{13}} = \frac{1 \cdot 143 \cdot 70 \cdot 1}{3^{10}} = \frac{10{,}010}{59{,}049}.$$

23. The probability that the committee contains exactly 3 faculty and 2 students is

$$\frac{C(7,3) \cdot C(6,2)}{C(13,5)} = \frac{35 \cdot 15}{1287} = \frac{175}{429}.$$

25. The number of arrangements in which the three diseased plants are adjacent is

$$\frac{6!}{1!\,5!}.$$

Thus the probability that the three diseased plants are all adjacent is

$$\frac{\left(\dfrac{6!}{1!\,5!}\right)}{\left(\dfrac{8!}{3!\,5!}\right)} = \frac{6}{56} = \frac{3}{28}.$$

27. The probability that Rebecca receives 2 books, Sheila receives 4 books, and Tom receives 3 books is

$$\frac{\left(\dfrac{9!}{2!\,4!\,3!}\right)}{3^9} = \frac{1260}{19{,}683} = \frac{140}{2187}.$$

29. As in Example 8.35, the probability that a randomly chosen permutation of the letters in the word "determine" has no adjacent e's is

$$\frac{P(6,6) \cdot C(7,3)}{\left(\dfrac{9!}{3!\,1!\,1!\,1!\,1!\,1!\,1!}\right)} = \frac{720 \cdot 35}{60{,}480} = \frac{5}{12}.$$

31. A hand containing one pair can be formed by choosing the denomination of the pair, picking two of the cards with that denomination, choosing the denominations of the other three cards, and then picking one card of each of these three denominations. Thus the number of different hands that contain one pair is

$$C(13,1) \cdot C(4,2) \cdot C(12,3) \cdot C(4,1) \cdot C(4,1) \cdot C(4,1) = 1{,}098{,}240.$$

Thus the probability of being dealt a hand containing one pair is

$$\frac{1{,}098{,}240}{C(52,5)} = \frac{1{,}098{,}240}{2{,}598{,}960} = \frac{352}{833}.$$

33. A hand containing three-of-a-kind can be formed by choosing the denomination of the three cards that are the same, picking the denominations of the other two cards, and then picking one card of each of these two denominations. Thus the number of different hands that contain three-of-a-kind is

$$C(13,1) \cdot C(4,3) \cdot C(12,2) \cdot C(4,1) \cdot C(4,1) = 54{,}912.$$

Thus the probability of being dealt a hand containing three-of-a-kind is

$$\frac{54{,}912}{C(52,5)} = \frac{54{,}912}{2{,}598{,}960} = \frac{88}{4165}.$$

35. Regard a selection of 5 of the 25 accounts as a 25-bit string containing five 1s and twenty 0s, where a 1 denotes a file that is chosen and a 0 denotes a file that is not chosen. An acceptable selection of the files is then a bit string of this form having no two consecutive 1s. As in Exercise 23 of Section 8.4, this number is $C(21,5)$. Hence the probability of selecting five files with no two having consecutive numbers is

$$\frac{C(21,5)}{C(25,5)} = \frac{20{,}349}{53{,}130} = \frac{969}{2530}.$$

8.6   THE PRINCIPLE OF INCLUSION-EXCLUSION

1. If B and F are the sets of moviegoers who liked films by Bergman and Fellini, respectively, then, by (8.4), the number of people who liked films by Bergman or Fellini is

$$|B \cup F| = |B| + |F| - |B \cap F| = 33 + 25 - 18 = 40.$$

3. Let R and G be the sets of union members who liked their representative and their governor, respectively. Then, by (8.4), the number of members who liked neither their representative nor their governor is

$$318 - |R \cup G| = 318 - (|R| + |G| - |R \cap G|)$$
$$= 318 - (127 + 84 - 53)$$
$$= 160.$$

5. Let C, M, and H denote the sets of residents who are college-educated, married, and home-owners, respectively. The number of residents who are not college-educated, not married, and not homeowners is

$$650 - |C \cup M \cup H| = 650 - [(310 + 356 + 328) - (180 + 147 + 166) + 94] = 55.$$

7. Let U denote the set of all fast-food restaurants in the city, and let H, B, and P denote the sets of restaurants that serve hamburgers, roast beef sandwiches, and pizza, respectively. Then $|U| - |H \cup B \cup P| = 5$. So the number of fast-food restaurants in the city is

$$|U| = |H \cup B \cup P| + 5 = [(13 + 8 + 10) - (5 + 3 + 2) + 1] + 5 = 27.$$

9. The number of ways to choose the four couples in which a wife and husband are partners is $C(8,4)$. Also, there are D_4 (the number of derangements of four objects) ways to assign partners to the other four wives so that no husband is assigned as a partner to his wife. Example 8.44 shows that $D_4 = 9$. Hence the probability that exactly four husbands are paired with their wives is

$$\frac{C(8,4) \cdot D_4}{P(8,8)} = \frac{70 \cdot 9}{40{,}320} = \frac{1}{64}.$$

11. Let A_i denote the set of dinners in which Alison ate with friend i, and let n_s be defined as in the principle of inclusion-exclusion. Then

$$n_1 = |A_1| + |A_2| + |A_3| + |A_4| + |A_5| + |A_6| + |A_7| = C(7,1) \cdot 15 = 7 \cdot 15 = 105,$$
$$n_2 = C(7,2) \cdot 8 = 21 \cdot 8 = 168,$$
$$n_3 = C(7,3) \cdot 6 = 35 \cdot 6 = 210,$$
$$n_4 = C(7,4) \cdot 5 = 35 \cdot 5 = 175,$$
$$n_5 = C(7,5) \cdot 4 = 21 \cdot 4 = 84,$$
$$n_6 = C(7,6) \cdot 3 = 7 \cdot 3 = 21,$$
$$n_7 = C(7,7) \cdot 0 = 0.$$

Thus Alison ate dinner with at least one of these friends on

$$|A_1 \cup A_2 \cup A_3 \cup A_4 \cup A_5 \cup A_6 \cup A_7| = n_1 - n_2 + n_3 - n_4 + n_5 - n_6 + n_7$$
$$= 105 - 168 + 210 - 175 + 84 - 21 + 0$$
$$= 35$$

occasions. Hence the number of occasions on which she did not eat with any of these friends is $6(7) - 35 = 7$.

13. Let U denote the set of all possible ways to assign colors to the vertices using at most k colors, A_1 denote the set of colorings in which V_1 and V_2 receive the same color, A_2 denote the set of colorings in which V_1 and V_3 receive the same color, A_3 denote the set of colorings in which V_2 and V_4 receive the same color, and A_4 denote the set of colorings in which V_3 and V_4 receive the same color. Then $|U| = k^4$ because there are k choices of color for each vertex, and $|A_1| = k^3$ since there are k choices of color for V_1, k choices of color for V_3, and k choices of color for V_4. Similar reasoning shows that $|A_2| = |A_3| = |A_4| = k^3$, and so

$$n_1 = |A_1| + |A_2| + |A_3| + |A_4| = 4k^3.$$

Furthermore,

$$|A_1 \cap A_2| = |A_1 \cap A_3| = |A_1 \cap A_4| = |A_2 \cap A_3| = |A_2 \cap A_4| = |A_3 \cap A_4| = k^2$$

because either three vertices must receive the same color or else two pairs of vertices must receive the same color, and so

$$n_2 = |A_1 \cap A_2| + |A_1 \cap A_3| + |A_1 \cap A_4| + |A_2 \cap A_3| + |A_2 \cap A_4| + |A_3 \cap A_4| = 6k^2.$$

Similar reasoning gives

$$n_3 = |A_1 \cap A_2 \cap A_3| + |A_1 \cap A_2 \cap A_4| + |A_1 \cap A_3 \cap A_4| + |A_2 \cap A_3 \cap A_4| = 4k,$$

and

$$n_4 = |A_1 \cap A_2 \cap A_3 \cap A_4| = k.$$

Thus the number of ways to assign one of at most k colors to the vertices so that no two adjacent vertices receive the same color is

$$\begin{aligned}
|U| - (n_1 - n_2 + n_3 + n_4) &= k^4 - (4k^3 - 6k^2 + 4k - k) \\
&= k^4 - 4k^3 + 6k^2 - 3k \\
&= k(k^3 - 4k^2 + 6k - 3) \\
&= k(k-1)(k^2 - 3k + 3).
\end{aligned}$$

15. If a positive integer less than 101 is divisible by a perfect square n^2, then $n \le 10$. Note that any positive integer that is divisible by 4^2, 6^2, 8^2, or 10^2 is divisible by 2^2 and that any positive integer divisible by 9^2 is divisible by 3^2. So it suffices to check for divisibility by 2^2, 3^2, 5^2, and 7^2. Let A_1, A_2, A_3, and A_4 denote the sets of positive integers less than 101 that are divisible by 2^2, 3^2, 5^2, and 7^2, respectively. Because there are $100/4 = 25$ positive integers less than 101 that are divisible by $2^2 = 4$, we see that $|A_1| = 25$. Likewise $|A_2| = 11$, $|A_3| = 4$, and $|A_4| = 2$. Elements of $A_1 \cap A_2$ are divisible by both 2^2 and 3^2 and hence are divisible by $2^2 \cdot 3^2 = 36$. Clearly the only positive integers less than 101 that are divisible by 36 are 36 and 72; thus $|A_1 \cap A_2| = 2$. Similar reasoning gives

$$|A_1 \cap A_3| = 1$$

and

$$
\begin{aligned}
|A_1 \cap A_4| &= |A_2 \cap A_3| = |A_2 \cap A_4| = |A_3 \cap A_4| \\
&= |A_1 \cap A_2 \cap A_3| = |A_1 \cap A_2 \cap A_4| = |A_1 \cap A_3 \cap A_4| = |A_2 \cap A_3 \cap A_4| \\
&= |A_1 \cap A_2 \cap A_3 \cap A_4| \\
&= 0.
\end{aligned}
$$

Hence

$$|A_1 \cup A_2 \cup A_3 \cup A_4| = (25 + 11 + 4 + 2) - (2 + 1) + 0 - 0 = 39.$$

So there are $100 - 39 = 61$ square free positive integers less than 101.

17. Let A_1, A_2, A_3, and A_4 denote the sets of positive integers less than 1000 that are crossed out in searching for multiples of 2, 3, 5, and 7, respectively, and let n_s be defined as in the principle of inclusion-exclusion. Because 2, 3, 5, and 7 are not crossed out in the list, we have

$$
\begin{aligned}
n_1 &= |A_1| + |A_2| + |A_3| + |A_4| \\
&= 499 + 332 + 199 + 141 \\
&= 1171.
\end{aligned}
$$

The numbers in the set $A_i \cap A_j$ are multiples of both i and j and so are multiples of ij. Hence

$$
\begin{aligned}
n_2 &= |A_1 \cap A_2| + |A_1 \cap A_3| + |A_1 \cap A_4| + |A_2 \cap A_3| + |A_2 \cap A_4| + |A_3 \cap A_4| \\
&= 166 + 100 + 71 + 66 + 47 + 28 \\
&= 478.
\end{aligned}
$$

Likewise

$$
\begin{aligned}
n_3 &= |A_1 \cap A_2 \cap A_3| + |A_1 \cap A_2 \cap A_4| + |A_1 \cap A_3 \cap A_4| + |A_2 \cap A_3 \cap A_4| \\
&= 33 + 23 + 14 + 9 \\
&= 79,
\end{aligned}
$$

and

$$n_4 = |A_1 \cap A_2 \cap A_3 \cap A_4| = 4.$$

Therefore the number of integers in the list that are crossed out as multiples of 2, 3, 5, or 7 is

$$n_1 - n_2 + n_3 - n_4 = 1171 - 478 + 79 - 4 = 768.$$

Hence the number of integers that are *not* crossed out is $999 - 768 = 231$.

19. For $i = 1$, 2, 3, and 4, let A_i denote the set of seating arrangements in which couple i is seated in adjacent seats. If couple 1 are seated in adjacent seats, then we regard couple 1 as a group of two. The number of ways to arrange this group and the six other individuals in a row of 8 seats is $P(7,7)$. Since there are 2 ways to arrange couple 1 (with the husband on the left or the wife on the left), we see that

$$|A_1| = 2 \cdot P(7,7) = 2 \cdot 5040 = 10{,}080.$$

Thus we have

$$n_1 = C(4,1) \cdot |A_1| = 4 \cdot 10{,}080 = 40{,}320.$$

Similar reasoning gives

$$|A_1 \cap A_2| = 2^2 \cdot P(6,6) = 4 \cdot 720 = 2880,$$

so that

$$n_2 = C(4,2) \cdot |A_1 \cap A_2| = 6 \cdot 2880 = 17{,}280.$$

Likewise

$$|A_1 \cap A_2 \cap A_3| = 2^3 \cdot P(5,5) = 8 \cdot 120 = 960,$$

and so

$$n_3 = C(4,3) \cdot |A_1 \cap A_2 \cap A_3| = 4 \cdot 960 = 3840.$$

Finally,

$$n_4 = |A_1 \cap A_2 \cap A_3 \cap A_4| = 2^4 \cdot P(4,4) = 16 \cdot 24 = 384.$$

Hence the number of seating arrangements in which some husband is seated beside his wife is

$$n_1 - n_2 + n_3 - n_4 = 40{,}320 - 17{,}280 + 3840 - 384 = 26{,}496.$$

Since the number of ways to arrange 8 persons in a row of 8 seats is $P(8,8)$, the number of seating arrangements in which no husband is seated beside his wife is

$$P(8,8) - (n_1 - n_2 + n_3 - n_4) = 40{,}320 - 26{,}496 = 13{,}824.$$

21. Let A_1, A_2, A_3, and A_4 denote the sets of all five-card poker hands that contain no spades, hearts, diamonds, and clubs, respectively. Then, as in Example 8.41,

$$|A_1| = C(39,5),$$
$$|A_1 \cap A_2| = C(26,5),$$
$$|A_1 \cap A_2 \cap A_3| = C(13,5),$$

and

$$|A_1 \cap A_2 \cap A_3 \cap A_4| = 0.$$

Therefore

$$|A_1 \cup A_2 \cup A_3 \cup A_4|$$
$$= C(4,1) \cdot C(39,5) - C(4,2) \cdot C(26,5) + C(4,3) \cdot C(13,5) - C(4,4) \cdot 0$$
$$= 4 \cdot 575{,}757 - 6 \cdot 65{,}780 + 4 \cdot 1287$$
$$= 1{,}913{,}496.$$

Hence the number of hands containing at least one card in each suit is

$$C(52,5) - |A_1 \cup A_2 \cup A_3 \cup A_4| = 2{,}598{,}960 - 1{,}913{,}496 = 685{,}464.$$

23. Let A_i denote the set of nonnegative integer solutions to the equation

$$x_1 + x_2 + x_3 + x_4 = 12$$

in which $x_i \geq 5$. Then

$$|A_i| = C(7 + 4 - 1, 7) = C(10,7)$$

because we can put five 1s into x_i and distribute the remaining seven 1s into any x_j. Likewise if $i \neq j$,

$$|A_i \cap A_j| = C(2 + 4 - 1, 2) = C(5,2).$$

Moreover, if i, j, and k are all distinct, then $A_i \cap A_j \cap A_k = \varnothing$ since otherwise $x_i + x_j + x_k \geq 15$. Thus

$$|A_1 \cup A_2 \cup A_3 \cup A_4| = C(4,1) \cdot C(10,7) - C(4,2) \cdot C(5,2) + 0 - 0$$
$$= 4(120) - 6(10)$$
$$= 420.$$

Hence the number of nonnegative integer solutions to the equation

$$x_1 + x_2 + x_3 + x_4 = 12$$

in which no x_i exceeds 4 is

$$C(12 + 4 - 1, 12) - 420 = C(15,12) - 420 = 455 - 420 = 35.$$

25. Let $S = \{a_1, a_2, \ldots, a_m\}$. Let U denote the set of all lists of length n chosen with repetition from S and, for $i = 1, 2, \ldots, m$, let A_i denote the set of all the lists in U in which a_i does not appear. Then $\overline{A_1} \cap \overline{A_2} \cap \cdots \cap \overline{A_m}$ is the set of all the lists in U in which each element of S appears at least once. Let n_s be as defined in the principle of inclusion-exclusion. Then

$$|\overline{A_1} \cap \overline{A_2} \cap \cdots \cap \overline{A_m}| = |U| - \left[n_1 - n_2 + \ldots + (-1)^{m-1} n_m \right].$$

Each permutation in A_i is a permutation of length n chosen with repetition from the set $S - \{a_i\}$. There are $(m-1)^n$ such permutations, and so $|A_i| = (m-1)^n$. Hence

$$n_1 = C(m,1) \cdot (m-1)^n.$$

Similarly, for $i \neq j$, each list in $A_i \cap A_j$ is a list of length n chosen with repetition from the set $S - \{a_i, a_j\}$. Thus $|A_i \cap A_j| = (m-2)^n$, and so

$$n_2 = C(m,2) \cdot (m-2)^n.$$

Likewise for any $s = 1, 2, \ldots, m$, we see that

$$n_s = C(m,s) \cdot (m-s)^n.$$

Therefore the number of permutations of elements of S in which each element of S appears at least once is

$$|U| - \left[n_1 - n_2 + \cdots + (-1)^{m-1} n_m \right]$$

$$= |U| - n_1 + n_2 - \ldots - (-1)^{m-1} n_m$$

$$= C(m,0)(m-0)^n - C(m,1)(m-1)^n + \ldots + (-1)^m C(m,m)(m-m)^n$$

$$= C(m,0)(m-0)^n - C(m,1)(m-1)^n + \ldots + (-1)^{m-1} C(m,m-1)(1)^n + 0$$

$$= C(m,0)(m-0)^n - C(m,1)(m-1)^n + \ldots + (-1)^{m-1} C(m,m-1)(1)^n.$$

27. Let U denote the set of all permutations of $1, 2, \ldots, k$ and A_i denote the set of all the permutations in U having an i in position i. Then $\overline{A_1} \cap \overline{A_2} \cap \cdots \cap \overline{A_k}$ is the set of all derangements of $1, 2, \ldots, k$. In A_i, the integer i occurs in position i and the remaining $k-1$ integers may occur in any position. Thus $|A_i| = (k-1)!$. So if n_s is as defined in the principle of inclusion-exclusion, then

$$n_1 = C(k,1) \cdot (k-1)! = k \cdot (k-1)! = \frac{k!}{1!}.$$

Similar reasoning shows that

$$n_s = C(k,s) \cdot (k-s)! = \frac{k!}{s!}$$

for $s = 1, 2, \ldots, k$. Thus

$$|\overline{A_1} \cap \overline{A_2} \cap \cdots \overline{A_k}| = |U| - \left[n_1 - n_2 + \cdots + (-1)^{k-1} n_k \right]$$

$$= k! \left[\frac{1!}{0!} - \frac{1}{1!} + \ldots + (-1)^k \frac{1}{k!} \right].$$

173

29. Clearly there are no ways to distribute n balls into 0 urns; so $S(n, 0) = 0$. Likewise there is only 1 way to distribute n balls into 1 urn, namely by putting all the balls into the single urn. Hence $S(n, 1) = 1$.

A distribution of n distinguishable balls into two indistinguishable urns leaving neither urn empty is simply a partition of the set of balls into two subsets. The number of such distributions is therefore $2^{n-1} - 1$ by Exercise 27 of Section 2.2; hence

$$S(n, 2) = 2^{n-1} - 1.$$

To distribute n balls into $n - 2$ urns leaving no urn empty, we must either put two balls into two urns and one ball into each of the others or put three balls into one urn and one ball into each of the others. In the first case, the four balls to be put into the two urns can be chosen in $C(n, 4)$ ways, and these four balls can be partitioned into two pairs in 3 ways. Thus there are $3 \cdot C(n, 4)$ distributions of the first type. In the second case, there are $C(n, 3)$ ways to choose the three balls to be put into the same urn. Hence

$$S(n, n - 2) = 3 \cdot C(n, 4) + C(n, 3).$$

To distribute n balls into $n - 1$ urns leaving no urn empty, we must put two balls into one urn and one ball into each of the others. If the urns are indistinguishable, two such distributions are distinct if and only if the urns containing two balls have different contents. But the number of different ways to choose two balls from among n distinguishable balls is $C(n, 2)$; thus

$$S(n, n - 1) = C(n, 2).$$

Finally, the only way to distribute n balls into n urns leaving no urn empty is by putting one ball into each urn. If the urns are indistinguishable, there is only one such distribution; so $S(n, n) = 1$.

31. Let X be a set containing exactly n elements. A partition of X into k subsets can be regarded as a distribution of the n (distinguishable) elements of X into k indistinguishable urns. Hence there are $S(n, k)$ partitions of X into k subsets. It follows that there are

$$S(n, 1) + S(n, 2) + \cdots + S(n, n)$$

partitions of X. By Theorem 2.4, this is the number of equivalence relations on X.

33. In any distribution of $n + 1$ distinguishable balls into m indistinguishable urns that leaves no urn empty, ball $n + 1$ is either the only ball in some urn or accompanied by other balls.

Case 1: Ball n + 1 occurs alone in some urn.
In this case, balls $1, 2, \ldots, n$ form a distribution of n balls into $m - 1$ indistinguishable urns leaving no urn empty. The number of such distributions is $S(n, m - 1)$.

Case 2: Ball $n + 1$ does not occur alone in some urn.
In this case, the given distribution can be obtained as follows. First distribute balls $1, 2, \ldots, n$ into all m urns leaving no urn empty. This distribution can be made in $S(n, m)$ ways. Then pick an urn for ball $n + 1$ in one of m ways. By the multiplication principle, the number of ways to perform both of these steps is $m \cdot S(n, m)$.

Combining cases 1 and 2 by the addition principle, we see that the number of ways to distribute $n + 1$ distinguishable balls into m indistinguishable urns that leaves no urn empty is

$$S(n, m - 1) + m \cdot S(n, m).$$

35. We will first count the number of ways to distribute n distinguishable balls into m *distinguishable* urns, leaving no urn empty. Since the urns are distinguishable, we may number them $1, 2, \ldots, m$. A distribution of the balls can be accomplished by choosing, for each of the n balls, the number of the urn into which it will be placed. Moreover, if we insist that each of the numbers $1, 2, \ldots, m$ be chosen at least once, then no urn will be left empty. Thus each distribution of n distinguishable balls into m distinguishable urns that leaves no urn empty is a list of length n chosen with repetition from a set of m elements. By Exercise 25, the number of such lists is

$$C(m, 0)(m - 0)^n - C(m, 1)(m - 1)^n + \ldots + (-1)^{m-1} C(m, m - 1)(1)^n.$$

Therefore this expression gives the number of ways to distribute n distinguishable balls into m distinguishable urns, leaving no urn empty. Now each distribution of n distinguishable balls into m *indistinguishable* urns that leaves no urn empty gives rise to $m!$ distributions of n distinguishable balls into m *distinguishable* urns that leaves no urn empty, namely the ones obtained by permuting the contents of the m urns in all $P(m, m) = m!$ ways. Hence

$$S(n, m) = \frac{1}{m!} \left[C(m, 0)(m - 0)^n - C(m, 1)(m - 1)^n + \ldots + (-1)^{m-1} C(m, m - 1)(1)^n \right].$$

37. Let $x \in S$. If x belongs to fewer than s of the subsets A_i, then x is not counted in any of the numbers $p_s, n_s, n_{s+1}, \ldots, n_r$. Hence both sides of the equation to be proved equal 0 in this case. If x belongs to precisely s of the subsets A_i, then x is counted once in p_s and n_s, but x is not counted in $n_{s+1}, n_{s+2}, \ldots, n_r$. So both sides of the equation equal 1 in this case. It remains to consider the case in which x belongs to precisely m of the subsets A_i, where $s < m \leq r$. In this case, x is counted in neither p_s nor any term n_k for $k > m$. However, x is counted $C(m, s)$ times in n_s, $C(m, s+1)$ times in n_{s+1}, $\ldots$, and $C(m, m)$ times in n_m. Thus, on the right side of the equation to be proved, x is counted

$$C(s, 0) \cdot C(m, s) - C(s + 1, 1) \cdot C(m, s + 1) + \cdots + (-1)^{m-s} C(m, m - s) \cdot C(m, m)$$

times. Since $p_s = 0$, it suffices to prove that this number is 0. Using (8.1), we have

175

$C(s + k, k) \cdot C(m, s + k) = C(m, s) \cdot C(m - s, k)$. Thus, by the binomial theorem,

$$C(s, 0) \cdot C(m, s) - C(s + 1, 1) \cdot C(m, s + 1) + \cdots + (-1)^{m-s} C(m, m - s) \cdot C(m, m)$$

$$= C(m, s) \cdot C(m - s, 0) - C(m, s) \cdot C(m - s, 1) + \cdots$$
$$+ (-1)^{m-s} C(m, s) \cdot C(m - s, m - s)$$

$$= C(m, s)[C(m - s, 0) - C(m - s, 1) + \cdots + (-1)^{m-s} C(m - s, m - s)]$$

$$= C(m, s)[(1 - 1)^{m-s}] = 0,$$

completing the proof.

8.7 GENERATING PERMUTATIONS AND r-COMBINATIONS

1. Since the first entry of p is less than the first entry of q, $p < q$.

3. Since the first entries of p and q are the same, but the second entry of p is less than the second entry of q, $p < q$.

5. Since the first two entries of p and q are the same, but the third entry of p is greater than the third entry of q, $p > q$.

7. Scanning p from right-to-left, the first entry of p that is less than the entry to its right is 5. Since there is only one entry of p to the right of 5, we interchange the last two entries of p to obtain its successor, which is $(2, 1, 4, 3, 6, 5)$.

9. Scanning p from right-to-left, the first entry of p that is less than the entry to its right is 4. Again scanning p from right-to-left, the rightmost entry of p that is greater than 4 is 5. Interchange the entries 4 and 5 in p to obtain $(2, 1, 5, 6, 4, 3)$. Then reverse the order of the entries to the right of 5 to obtain the successor of p, which is $(2, 1, 5, 3, 4, 6)$.

11. As in Exercise 9, the successor of p is $(5, 6, 4, 1, 2, 3)$.

13. Since no entry of $(6, 5, 4, 3, 2, 1)$ is less than the entry to its right, the permutation $(6, 5, 4, 3, 2, 1)$ has no successor.

15. As in Exercise 9, the successor of p is $(5, 3, 1, 2, 4, 6)$.

17. As in Exercise 9, the successor of p is $(6, 4, 1, 2, 3, 5)$.

19. By applying the algorithm for the lexicographic ordering of permutations with $n = 4$, we obtain the following ordering of the permutations of $\{1, 2, 3, 4\}$:

$(1, 2, 3, 4)$; $(1, 2, 4, 3)$; $(1, 3, 2, 4)$; $(1, 3, 4, 2)$; $(1, 4, 2, 3)$; $(1, 4, 3, 2)$;
$(2, 1, 3, 4)$; $(2, 1, 4, 3)$; $(2, 3, 1, 4)$; $(2, 3, 4, 1)$; $(2, 4, 1, 3)$; $(2, 4, 3, 1)$;
$(3, 1, 2, 4)$; $(3, 1, 4, 2)$; $(3, 2, 1, 4)$; $(3, 2, 4, 1)$; $(3, 4, 1, 2)$; $(3, 4, 2, 1)$;
$(4, 1, 2, 3)$; $(4, 1, 3, 2)$; $(4, 2, 1, 3)$; $(4, 2, 3, 1)$; $(4, 3, 1, 2)$; $(4, 3, 2, 1)$.

21. To find the successor of the subset $S = \{1, 3, 5, 7, 9\}$ in the lexicographic ordering of the 5-element subsets of $\{1, 2, 3, 4, 5, 6, 7, 8, 9\}$, we find the rightmost element in S that does not equal its maximum value. This is the element 7, which we increase to 8. Since every element to the right of 8 is as small as possible, no other changes are necessary. Thus the successor of S is $\{1, 3, 5, 8, 9\}$.

23. To find the successor of the subset $S = \{2, 3, 5, 8, 9\}$ in the lexicographic ordering of the 5-element subsets of $\{1, 2, 3, 4, 5, 6, 7, 8, 9\}$, we proceed as in Exercise 21. The rightmost element in S that does not equal its maximum value is the element 5, which we increase to 6. We then replace the elements to the right of 6 by $6 + 1 = 7$ and $7 + 1 = 8$. Thus the successor of S is $\{2, 3, 6, 7, 8\}$.

25. As in Exercise 21, the successor of $S = \{3, 4, 5, 7, 8\}$ in the lexicographic ordering of the 5-element subsets of $\{1, 2, 3, 4, 5, 6, 7, 8, 9\}$ is $\{3, 4, 5, 7, 9\}$.

27. To find the successor of the subset $S = \{3, 5, 7, 8, 9\}$ in the lexicographic ordering of the 5-element subsets of $\{1, 2, 3, 4, 5, 6, 7, 8, 9\}$, we proceed as in Exercise 21. Thus the successor of S is $\{3, 6, 7, 8, 9\}$.

29. As in Exercise 21, the successor is $\{4, 6, 7, 8, 9\}$.

31. Since every element is as large as possible, $\{5, 6, 7, 8, 9\}$ has no successor.

SUPPLEMENTARY EXERCISES

1. $C(9, 7) = \dfrac{9!}{7!\,(9 - 7)!} = \dfrac{9 \cdot 8 \cdot 7!}{7!\,2!} = \dfrac{9 \cdot 8}{2 \cdot 1} = 36$

3. $P(9, 4) = \dfrac{9!}{(9 - 4)!} = \dfrac{9 \cdot 8 \cdot 7 \cdot 6 \cdot 5!}{5!} = 9 \cdot 8 \cdot 7 \cdot 6 = 3024$

5. The binomial theorem gives $(x - 1)^6 = x^6 - 6x^5 + 15x^4 - 20x^3 + 15x^2 - 6x + 1$.

7. The binomial theorem gives

$$(2x + 3y)^5 = (2x)^5 + 5(2x)^4(3y) + 10(2x)^3(3y)^2 + 10(2x)^2(3y)^3 + 5(2x)(3y)^4 + (3y)^5$$
$$= 32x^5 + 240x^4 y + 720x^3 y^2 + 1080x^2 y^3 + 810xy^4 + 243y^5.$$

9. Scanning $p = (8, 2, 3, 7, 6, 5, 4, 1)$ from right-to-left, the first entry of p that is less than the entry to its right is 3. Again scanning p from right-to-left, the rightmost entry of p that is greater than 3 is 4. Interchange the entries 3 and 4 in p to obtain

$$(8, 2, 4, 7, 6, 5, 3, 1).$$

Then reverse the order of the entries to the right of 4 to obtain the successor of p, which is

$$(8, 2, 4, 1, 3, 5, 6, 7).$$

11. By Theorem 8.2, the last nine numbers in the $n = 17$ row of Pascal's triangle are the first nine numbers in the row written in reverse order. Thus these numbers are: 24310, 19448, 12376, 6188, 2380, 680, 136, 17, and 1.

13. By Theorem 8.1, $C(18, 12) = C(17, 12) + C(17, 11) = 6188 + 12{,}376 = 18{,}564.$

15. According to the addition principle, the number of ways to select one snack is

$$5 + 12 + 4 = 21.$$

17. There are 6 ways to select the slacks, 8 ways to select the blouse, 5 ways to select the shoes, and 4 ways to select the purse. (There are three purses and the option of taking no purse.) Hence the multiplication principle gives the number of outfits to be $6 \cdot 8 \cdot 5 \cdot 4 = 960.$

19. Since the order of performance matters, this is a problem involving permutations. The number of different programs is $P(14, 5) = 240{,}240.$

21. In order to obtain a pair of integers with an even sum, we must select either two odd integers or two even integers. The probability of doing this is

$$\frac{C(30, 2) + C(30, 2)}{C(60, 2)} = \frac{2 \cdot 435}{1770} = \frac{29}{59}.$$

23. The number of different ways to choose the students is given by Theorem 8.7 to be $C(12, 3) \cdot C(9, 4) \cdot C(5, 5) = 220 \cdot 126 \cdot 1 = 27{,}720.$

25. The word "rearrangement" has two a's, three e's, one g, one m, two n's, three r's, and one t. Theorem 8.7 gives the number of distinguishable rearrangements of its letters to be

$$\frac{13!}{2!\,3!\,1!\,1!\,2!\,3!\,1!} = 43{,}243{,}200.$$

27. The number of selections is $C(16, 4) = 1820.$

29. The addition principle gives the number of selections as $4 + 7 + 3 = 14.$

31. (a) The number of possible 5-digit numbers using the digits 1–7 without repetition is $P(7,5) = 2520$.

(b) If the number must begin with 6, then there is only 1 choice for the first digit, 6 choices for the second digit, 5 choices for the third, 4 choices for the fourth, and 3 choices for the fifth. Hence there are $1 \cdot 6 \cdot 5 \cdot 4 \cdot 3 = 360$ such numbers. Thus the probability of randomly selecting a number that begins with 6 is

$$\frac{360}{2520} = \frac{1}{7}.$$

(c) There are $P(5,2)$ positions for the digits 1 and 2. The remaining three digits in the number can be chosen and arranged in $P(5,3)$ ways. Hence there are $P(5,2) \cdot P(5,3) = 20 \cdot 60 = 1200$ numbers that contain the digits 1 and 2. The probability of selecting such a number is therefore

$$\frac{1200}{2520} = \frac{10}{21}.$$

33. (a) The number of ways to select four different officers is $P(30,4) = 657{,}720$.

(b) The number of ways to select a five-member executive committee is $C(30,5) = 142{,}506$.

35. The number of recipients selected can be 0, 1, 2, 3, 4, or 5. Thus the number of different ways to select the recipients is

$$C(6,0) + C(6,1) + C(6,2) + C(6,3) + C(6,4) + C(6,5),$$

which is the sum of the first six entries in the $n = 6$ row of Pascal's triangle. This number is $1 + 6 + 15 + 20 + 15 + 6 = 63$.

37. As in Exercise 35 of Section 8.5, we will regard a selection of 10 of the integers $1, 2, \ldots, 40$ as a 40-bit string containing ten 1s and thirty 0s. (Here a 1 denotes an integer that is selected, and a 0 denotes an integer that is not selected.) To obtain a selection containing no two consecutive integers, we arrange the thirty 0s in a row and choose positions for the ten 1s from among the 31 locations before the first 0, after the last 0, and between consecutive 0s. The number of such selections is $C(31,10)$. Hence the probability of choosing ten of the integers $1, 2, \ldots, 40$ with no two consecutive is

$$\frac{C(31,10)}{C(40,10)} = \frac{44{,}352{,}165}{847{,}660{,}528} = \frac{10{,}005}{191{,}216}.$$

39. Choose two cupcakes of each type. The remaining 12 cupcakes can be selected in $C(12 + 6 - 1, 12) = C(17, 12) = 6188$ different ways.

41. There are 90,000 positive integers between 10,000 and 99,999 (inclusive). The number of these that do not contain a zero is $9 \cdot 9 \cdot 9 \cdot 9 \cdot 9 = 59{,}049$. Hence the number of such integers that do contain a zero is $90{,}000 - 59{,}049 = 30{,}951$. The probability of selecting one of these at random is therefore

$$\frac{30{,}951}{90{,}000} = \frac{3439}{10{,}000}.$$

43. By Theorem 8.7, the number of distinguishable arrangements is

$$\frac{18!}{4!\,6!\,3!\,5!} = 514{,}594{,}080.$$

45. By Theorem 8.7, the number of ways to assign the drugs is

$$C(16,4) \cdot C(12,4) \cdot C(8,4) \cdot C(4,4) = 1820 \cdot 495 \cdot 70 \cdot 1 = 63{,}063{,}000.$$

47. By Theorem 8.7, the number of distinguishable arrangements is

$$\frac{15!}{4!\,6!\,5!} = 630{,}630.$$

49. Consider an arrangement of the integers $1, 2, \ldots, 12$ around a circle. For $k = 1, 2, \ldots, 12$, let S_k denote the set of five consecutive integers in this arrangement that starts with the integer k. Since each integer lies in five different sets S_k, the sum of the numbers in all of the sets S_k is

$$5(1 + 2 + \cdots + 12) = 5 \cdot \frac{12 \cdot 13}{2} = 390.$$

Now $390 > 384 = 12(32)$. Hence if we place 390 ones into twelve sets, the pigeonhole principle guarantees that some S_k must contain at least 33 ones. That is, some five consecutive integers must have a sum of 33 or more.

51. Let A_1, A_2, A_3, and A_4 denote the sets of all twelve-card pinochle hands that contain no ace of spades, no ace of hearts, no ace of diamonds, and no ace of clubs, respectively. Let n_s be defined as in the principle of inclusion-exclusion. The hands in A_1 consist of 12 cards chosen from the 46 pinochle cards other than the aces of spades; thus $|A_1| = C(46, 12)$, and so

$$n_1 = C(4,1) \cdot |A_1| = C(4,1) \cdot C(46,12).$$

Similarly

$$|A_1 \cap A_2| = C(44,12), \qquad |A_1 \cap A_2 \cap A_3| = C(42,12),$$

and

$$|A_1 \cap A_2 \cap A_3 \cap A_4| = C(40,12).$$

Hence

$$n_2 = C(4,2) \cdot C(44,12),$$

$$n_3 = C(4,3) \cdot C(42,12),$$

and

$$n_4 = C(4,4) \cdot C(40,12).$$

Therefore the number of hands that do not contain an ace of each suit is

$$|A_1 \cap A_2 \cap A_3 \cap A_4| = n_1 - n_2 + n_3 - n_4$$

$$= 4 \cdot C(46,12) - 6 \cdot C(44,12) + 4 \cdot C(42,12) - 1 \cdot C(40,12)$$

$$= 4(38{,}910{,}617{,}655) - 6(21{,}090{,}682{,}613)$$
$$+ 4(11{,}058{,}116{,}888) - 5{,}586{,}853{,}480$$

$$= 67{,}743{,}989{,}014,$$

and so the number of hands that contain an ace of each suit is

$$C(48,12) - 67{,}743{,}989{,}014 = 69{,}668{,}534{,}468 - 67{,}743{,}989{,}014 = 1{,}924{,}545{,}454.$$

Hence the probability that a random hand contains an ace of each suit is

$$\frac{1{,}924{,}545{,}454}{C(48,12)} = \frac{1{,}924{,}545{,}454}{69{,}668{,}534{,}468} = \frac{105{,}293}{3{,}811{,}606}.$$

53. If $0 \leq k \leq n$, then a set with n elements has $C(n,k)$ subsets containing k elements and 2^n subsets of any size. Thus $C(n,k) \leq 2^n$.

55. The sum to be evaluated is like the one in Exercise 35 of Section 8.3. We will prove by induction that for $n \geq 2$,

$$C(2,2) + C(3,2) + \cdots + C(n,2) = C(n+1,3).$$

For $n = 2$, the left side is $C(2,2) = 1$ and the right side is $C(3,3) = 1$; so the equation is correct for $n = 2$. Assume that the equation is correct for $n = k$, that is, assume

$$C(2,2) + C(3,2) + \cdots + C(k,2) = C(k+1,3).$$

Then, by the induction hypothesis and Theorem 8.1, we have

$$C(2,2) + C(3,2) + ... + C(k,2) + C(k+1,2) = C(k+1,3) + C(k+1,2)$$
$$= C(k+2,3),$$

which proves the equation for $n = k+1$. Therefore the equation is correct for all integers $n \geq 2$ by the principle of mathematical induction.

57. The terms in the expansion of $(x_1 + x_2 + \cdots + x_k)^n$ are of the form $x_1^{n_1} x_2^{n_2} \cdots x_k^{n_k}$, where the sum of the exponents is n. Such a term is obtained by choosing x_1 from

n_1 factors of $(x_1 + x_2 + \cdots + x_k)$, x_2 from n_2 factors, ..., x_k from n_k factors. By Theorem 8.7, the number of such choices is

$$\frac{n!}{n_1! \, n_2! \cdots n_k!}.$$

Hence the coefficient of the term $x_1^{n_1} x_2^{n_2} \cdots x_k^{n_k}$ in the expansion of $(x_1 + x_2 + \cdots + x_k)^n$ is

$$\frac{n!}{n_1! \, n_2! \cdots n_k!}.$$

59. Consider the sequence of sets $\varnothing, \{1\}, \{1,2\}, \{1,2,3\}, \ldots, \{1,2,\ldots,n\}$. Since this sequence contains $n+1$ sets, the pigeonhole principle guarantees that at least two of them, say A and B, where $A \subseteq B$, must be assigned the same color. Then $A \cup B = B$ and $A \cap B = A$, and so the sets A, B, $A \cup B$, and $A \cap B$ are all assigned the same color.

If there are $n+1$ colors available, this conclusion need not be true. For example, assign to each set X the color $|X| + 1$. If $|A| = |B|$ but $A \neq B$, then we have $|A \cap B| < |A| < |A \cup B|$. Hence for this assignment of colors, it is not the case that the four sets A, B, $A \cup B$, and $A \cap B$ are assigned the same color unless $A = B$.

Chapter 9

Recurrence Relations and Generating Functions

9.1   RECURRENCE RELATIONS

1. We begin with the initial condition:
$$s_0 = 5.$$

   Now, using the recurrence relation $s_n = 3s_{n-1} - 9$, we obtain the subsequent values:

$$s_1 = 3s_0 - 9 = 3(5) - 9 = 6,$$
$$s_2 = 3s_1 - 9 = 3(6) - 9 = 9,$$
$$s_3 = 3s_2 - 9 = 3(9) - 9 = 18,$$
$$s_4 = 3s_3 - 9 = 3(18) - 9 = 45,$$
$$s_5 = 3s_4 - 9 = 3(45) - 9 = 126.$$

3. Proceeding as in Exercise 1, we obtain the values $5, 13, 32, 73, 158, 331$.

5. Proceeding as in Exercise 1, we obtain the values $2, -3, -4, -11, -26, -63$.

7. Proceeding as in Exercise 1, we obtain the values $3, 4, 1, 10, -7, 56$.

9. Proceeding as in Exercise 1, we obtain the values $2, -1, 4, 5, 15, 31$.

11. Proceeding as in Exercise 1, we obtain the values $1, 2, 5, 3, 13, 3$.

13. Since the tuition for the next year can be found by adding the previous year's amount and the 5.25 percent increase, we can multiply the previous year's tuition by 1.0525 to get the next year's tuition. The initial value is 28,000 dollars. Thus the recurrence relation and initial condition are $t_n = 1.0525t_{n-1}$ for $n \geq 1$ and $t_0 = 28,000$.

15. Since the number of franchises in each year is 6 more than the number of franchises in the previous year, we add 6 to the number from the previous year. The initial value is 24 franchises. Thus the recurrence relation and initial condition are: $r_n = r_{n-1} + 6$ for $n \geq 1$ and $r_0 = 24$.

17. To obtain the next month's balance, we first increase the previous month's balance by 1.5 percent, or equivalently, multiply the preceding month's balance by 1.015. From this amount, we subtract the $25 monthly payment to get the balance for the next month. Because the initial value is $280, the recurrence relation and initial condition are: $b_n = 1.015 b_{n-1} - 25$ for $n \geq 1$ and $b_0 = 280$.

19. Since 85 percent of the waste in the core room is eliminated each week, 15 percent of it remains. Thus the amount for the next week is found by multiplying the remaining amount from the preceding week by 0.15 and adding 2 kg for the new waste generated. The initial value is 1.7 kg. Thus the recurrence relation and initial condition are: $w_n = 0.15 w_{n-1} + 2$ for $n \geq 1$ and $w_0 = 1.7$.

21. This problem is similar to Example 9.3. First, note that there are 2 ways to spend 1 dollar (buy tape or buy a ruler). There are 7 sequences in which 2 dollars can be spent—3 involving a single purchase of 2 dollars and $2 \cdot 2 = 4$ involving two purchases of 1 dollar each. Similar reasoning shows that there are 21 sequences in which 3 dollars can be spent—1 involving a single purchase costing 3 dollars, $3 \cdot 2 = 6$ involving purchases of 2 dollars on the first day and three dollars on the second day, another $2 \cdot 3 = 6$ involving purchases of 1 dollar on the first day and 2 dollars on the second day, and $2 \cdot 2 \cdot 2 = 8$ involving purchases of 1 dollar on three successive days. Thus $s_1 = 2, 2_2 = 7$, and $s_3 = 21$.

In general, we must consider the ways that n dollars could have been spent. If the last item purchased cost 3 dollars, then there is only one possible purchase (a binder) and s_{n-3} ways of spending the previous $n - 3$ dollars. Thus there are s_{n-3} ways to spend n dollars with a last purchase of 3 dollars. Because there are three items costing 2 dollars, there are 3 ways to spend 2 dollars and s_{n-2} ways to spend the preceding $n-2$ dollars. Thus there are $3 s_{n-2}$ ways to spend n dollars with a last purchase of 2 dollars. Similarly, there are $2 s_{n-1}$ ways to spend n dollars with a last purchase of 1 dollar. Thus the recurrence relation and initial conditions are: $s_n = 2 s_{n-1} + 3 s_{n-2} + s_{n-3}$ for $n \geq 4$ and $s_1 = 2, s_2 = 7, s_3 = 21$.

23. Consider a sequence of the integers $1, 2, \ldots, n$. There are n choices for the last integer in the sequence and a_{n-1} arrangements of the preceding integers. Thus the recurrence relation and initial condition are: $a_n = n a_{n-1}$ for $n \geq 2$ and $a_1 = 1$.

25. The number of two-element subsets of a set $\{e_1, e_2, \ldots, e_{n-1}\}$ with $n - 1$ elements is s_{n-1}. If a new element, e_n, is available, all the new two-element subsets are formed by pairing e_n with each of the elements $e_1, e_2, \ldots, e_{n-1}$ to form a new two-element subset. This creates $n - 1$ new two-element subsets. Hence the recurrence relation and initial condition are: $s_n = s_{n-1} + (n - 1)$ for $n \geq 1$ and $s_0 = 0$.

27. This exercise is solved in a manner similar to Example 9.3. The initial conditions are $s_1 = 1$ (a single nickel), $s_2 = 2$ (two nickels or one dime), $s_3 = 3$ (three nickels, a nickel followed by a dime, or a dime followed by a nickel), and similarly, $s_4 = 5$, and $s_5 = 9$.

In general, suppose that $5n$ cents has been inserted into the vending machine, and consider the last coin used in the purchase. If the last coin is a quarter, the value of the previous coins is $5(n-5)$ cents. The number of ways to insert these coins is therefore s_{n-5}. Thus there are s_{n-5} ways to insert the coins if the last coin is a quarter. In like manner, if the last coin is a dime or a nickel, then there are s_{n-2} and s_{n-1} ways, respectively, to insert the previous coins. Hence the recurrence relation is $s_n = s_{n-1} + s_{n-2} + s_{n-5}$ for $n \geq 6$. In all, there are 128 sequences in which the coins could be inserted to pay for a drink costing 50 cents.

29. The initial value is 1 because there is only 1 way to group two people into a single pair. In general, the number of groupings of $2n$ people into n pairs can be found by first pairing any individual with one of the remaining $2n-1$ people, and then grouping the remaining $2n-2 = 2(n-1)$ people into $n-1$ pairs. Since the number of groupings of $2(n-1)$ people into $n-1$ pairs is c_{n-1}, there are $(2n-1)c_{n-1}$ such groupings. Thus the recurrence relation and initial condition are: $c_n = (2n-1)c_{n-1}$ for $n \geq 2$ and $c_1 = 1$.

31. By drawing 2 intersecting circles, we see that $r_1 = 2$. In general, suppose that we have $n-1$ circles dividing the plane into r_{n-1} regions. An nth circle that is drawn according to the requirements of the exercise will intersect the original $n-1$ circles in $2(n-1)$ points. These points partition the nth circle into $2(n-1)$ arcs, each of which divides a different one of the original r_{n-1} regions in two, thereby creating $2(n-1)$ additional regions. Thus the recurrence relation and initial condition are $r_n = r_{n-1} + 2(n-1)$ for $n \geq 2$ and $r_1 = 2$.

33. Strings of length 1 or 2 cannot contain three consecutive zeros, and the only string of length 3 having three consecutive zeros is 000. Hence

$$s_1 = 2, \qquad s_2 = 2^2 = 4, \qquad \text{and} \qquad s_3 = 2^3 - 1 = 7.$$

Consider a string of length n that does not have three consecutive zeros. Such a string must end in no zeros, exactly one zero, or exactly two zeros, that is, it must end in 1, 10, or 100. If the last digit is 1, it may be preceded by any of the s_{n-1} strings of length $n-1$. If a string ends in 10, these digits may be preceded by any of the s_{n-2} strings of length $n-2$. Finally, if a string ends in 100, these digits may be preceded by any of the s_{n-3} strings of length $n-3$. Thus the recurrence relation is $s_n = s_{n-1} + s_{n-2} + s_{n-3}$ for $n \geq 4$ with initial conditions $s_1 = 2$, $s_2 = 4$, and $s_3 = 7$.

35. As in Example 9.5, we define $c_0 = 1$. For $n = 1$, the configuration involves 2 points and 1 chord. Clearly there is only one way to draw a single chord through 2 points; so $c_1 = 1$. Therefore the initial conditions are $c_0 = 1$ and $c_1 = 1$.

In general, suppose that we have $2n$ points on a circle that are joined by n nonintersecting chords. Number the points $1, 2, \ldots, 2n$ in a clockwise direction around the circle. We claim that point 1 must be joined by a chord to an even-numbered point m. Otherwise, there would be an odd number of points between points 1 and m, and so the chord through one of these points must join it to one of the points $m+1, m+2, \ldots, 2n$. Such a chord would intersect the chord joining points 1 and m. Thus m is even, and so $m = 2k$ for some $k = 1, 2, \ldots, n$.

The chord through points 1 and $2k$ therefore partitions the points other than 1 and $2k$ into two sets $\{2, 3, \ldots, 2k-1\}$ and $\{2k+1, 2k+2, \ldots, 2n\}$, which lie on opposite sides of the chord through points 1 and $2k$. Since none of the chords intersect, all the chords must join two points in the same set. The first of these sets contains $2k - 2 = 2(k - 1)$ points, and the second contains $2n - 2k = 2(n - k)$ points. Thus there are c_{k-1} possible ways to draw the chords joining points in the first set, and c_{n-k} possible ways to draw the chords joining points in the second set. Therefore the multiplication principle shows that the number of ways to draw n nonintersecting chords with point 1 joined to point $2k$ is $c_{k-1}c_{n-k}$. But this can occur for $k = 1, 2, \ldots, n$, and so the addition principle gives the recurrence relation

$$c_n = c_0 c_{n-1} + c_1 c_{n-2} + \cdots + c_{n-1} c_0.$$

37. We will call a string "acceptable" if it does not contain the pattern 010. For $n = 1$ and $n = 2$, no n-bit string contains the pattern 010, and for $n = 3$, the only string containing this pattern is 010. Hence $s_1 = 2^1 = 2$, $s_2 = 2^2 = 4$, and $s_3 = 2^3 - 1 = 7$.

For an integer $k \geq 2$, let u_k and v_k denote the number of acceptable k-bit strings that end in 0 and 1, respectively. Then $s_k = u_k + v_k$. Also, an acceptable k-bit string ending in 1 is just an acceptable $(k-1)$-bit string followed by 1; so $v_k = s_{k-1}$. Finally, consider an acceptable k-bit string that ends in 0. Such a string is an acceptable $(k-1)$-bit string followed by 0, except that we must exclude the acceptable $(k-2)$-bit strings that end in 0 and are followed by 10. Thus the number of acceptable k-bit strings that end in 0 is $u_k = s_{k-1} - u_{k-2}$. Hence

$$
\begin{aligned}
s_n &= u_n + v_n \\
&= (s_{n-1} - u_{n-2}) + s_{n-1} \\
&= 2s_{n-1} - u_{n-2} \\
&= 2s_{n-1} - (s_{n-2} - v_{n-2}) \\
&= 2s_{n-1} - s_{n-2} + v_{n-2} \\
&= 2s_{n-1} - s_{n-2} + s_{n-3}.
\end{aligned}
$$

To compute s_6, we iterate this recurrence relation using the initial conditions:

$$
\begin{aligned}
s_4 &= 2s_3 - s_2 + s_1 = 2(7) - 4 + 2 = 12 \\
s_5 &= 2s_4 - s_3 + s_2 = 2(12) - 7 + 4 = 21 \\
s_6 &= 2s_5 - s_4 + s_3 = 2(21) - 12 + 7 = 37.
\end{aligned}
$$

9.2   THE METHOD OF ITERATION

1. The statement holds for $n = 1$: $1^2 - 1 + 2 = 2 = r_1$. Suppose that it holds for $n = k \geq 1$, that is, $r_k = k^2 - k + 2$. Using the recurrence relation, we see that

$$
\begin{aligned}
r_{k+1} &= r_{(k+1)-1} + 2[(k+1) - 1] \\
&= r_k + 2(k+1) - 2 \\
&= (k^2 - k + 2) + 2k + 2 - 2 \\
&= k^2 + 2k - k + 2 \\
&= (k^2 + 2k + 1) - (k+1) + 2 \\
&= (k+1)^2 - (k+1) + 2.
\end{aligned}
$$

Hence the statement holds for $n = k + 1$ and so, by the principle of mathematical induction, for each positive integer n.

3. The statement holds for $n = 0$: $4^0 - 3^0 + 1 = 1 = s_0$. It also holds for $n = 1$: $4^1 - 3^1 + 1 = 2 = s_1$. Suppose that it holds for $n = 0, 1, 2, \ldots, k$ where $k \geq 1$. Using the recurrence relation, we see that

$$
\begin{aligned}
s_{k+1} &= 7s_k - 12s_{k-2} + 6 \\
&= 7(4^k - 3^k + 1) - 12(4^{k-1} - 3^{k-1} + 1) + 6 \\
&= 7(4^k - 3^k + 1) - 3(4^k) - 4(3^k) - 12 + 6 \\
&= 4(4^k) - 3(3^k) + 7 - 6 \\
&= 4^{k+1} - 3^{k+1} + 1.
\end{aligned}
$$

Hence the statement holds for $n = k+1$ and so, by the strong principle of mathematical induction, for each nonnegative integer n.

5. The statement holds for $n = 2$: $2! \cdot \frac{1}{2!} = 1 = D_2$. Suppose that it holds for $n = k \geq 2$, that is,

$$
D_k = k! \left[\frac{1}{2!} - \frac{1}{3!} + \frac{1}{4!} - \cdots + (-1)^k \frac{1}{k!} \right].
$$

Using the recurrence relation, we see that

$$
\begin{aligned}
D_{k+1} &= (k+1)D_k + (-1)^{k+1} \\
&= (k+1) \left(k! \left[\frac{1}{2!} - \frac{1}{3!} + \frac{1}{4!} - \cdots + (-1)^k \frac{1}{k!} \right] \right) + (-1)^{k+1} \\
&= (k+1)! \left[\frac{1}{2!} - \frac{1}{3!} + \frac{1}{4!} - \cdots + (-1)^k \frac{1}{k!} + (-1)^{k+1} \frac{1}{(k+1)!} \right].
\end{aligned}
$$

Hence the statement holds for $n = k + 1$ and so, by the principle of mathematical induction, for each integer $n \geq 2$.

7. The statement holds for $n = 2$ because

$$C(2(2) + 2, 3) = C(6, 3) = 20 \quad \text{and} \quad s_2 = s_1 + 4(2)^2 = 4 + 16 = 20.$$

Suppose that it holds for $n = k \geq 2$, that is, $s_k = C(2k + 2, 3)$. Using the recurrence relation, we see that

$$s_{k+1} = s_k + 4(k + 1)^2 = C(2k + 2, 3) + 4(k + 1)^2 = C(2(k + 1) + 2, 3).$$

Hence the statement holds for $n = k + 1$ and so, by the principle of mathematical induction, for each integer $n \geq 2$.

9. Let $x_n = 2^2 + 4^2 + \cdots + (2n)^2$ for $n \geq 1$. Then $x_n - x_{n-1} = (2n)^2 = 4n^2$ for $n \geq 2$. Thus $x_n = x_{n-1} + 4n^2$ for $n \geq 2$ and $x_1 = 4$. Hence, by Exercise 7, $x_n = C(2n + 2, 3)$, that is, $2^2 + 4^2 + \cdots + (2n)^2 = C(2n + 2, 3) = \frac{2}{3}n(2n + 1)(n + 1)$.

11. From the recurrence relation and the initial condition, we have

$$s_0 = 9$$
$$s_1 = 9 + 4$$
$$s_2 = (9 + 4) + 4$$
$$s_3 = [(9 + 4) + 4] + 4$$
$$\vdots$$

Thus we conjecture that $s_n = 9 + 4n$ is a solution to the recurrence relation for $n \geq 0$. This conjecture can be established by mathematical induction.

13. From the recurrence relation and the initial condition, we have

$$s_0 = 5$$
$$s_1 = 3 \cdot 5$$
$$s_2 = 3 \cdot (3 \cdot 5)$$
$$s_3 = 3 \cdot [3 \cdot (3 \cdot 5)]$$
$$\vdots$$

Thus we conjecture that $s_n = 5(3^n)$ is a solution to the recurrence relation for $n \geq 0$. This conjecture can be established by mathematical induction.

15. From the recurrence relation and the initial condition, we have

$$s_0 = 6$$
$$s_1 = -6 = 6(-1)^1$$
$$s_2 = -[6(-1)^1] = 6(-1)^2$$
$$s_3 = -[6(-1)^2] = 6(-1)^3$$
$$\vdots$$

Thus we conjecture that $s_n = 6(-1)^n$ is a solution to the recurrence relation for $n \geq 0$. This conjecture can be established by mathematical induction.

17. From the recurrence relation and the initial condition, we have

$$s_0 = 1$$
$$s_1 = 5(1) + 3 = 5 + 3$$
$$s_2 = 5[5(1) + 3] = 5^2 + 5(3)$$
$$s_3 = 5[5^2 + 5(3)] + 3 = 5^3 + 5^2(3) + 3$$
$$s_4 = 5[5^3 + 5^2(3) + 3] + 3 = 5^4 + 5^3(3) + 5^2(3) + 5(3) + 3$$
$$\vdots$$

Thus we conjecture that

$$s_n = 5^n + 5^{n-1}(3) + \cdots + 5(3) + 3 = 5^n + 3(5^{n-1} + \cdots + 5 + 1)$$

is a solution to the recurrence relation for $n \geq 0$. Using the formula from Example 2.51, we can write this conjecture more compactly as

$$s_n = 5^n + 3(5^{n-1} + \cdots + 5 + 1)$$
$$= 5^n + 3\left(\frac{5^n - 1}{5 - 1}\right)$$
$$= 5^n + \frac{3}{4}(5^n - 1)$$
$$= 1.75(5^n) - 0.75.$$

This conjecture can be established by mathematical induction.

189

19. From the recurrence relation and the initial condition, we have

$$s_0 = 10$$
$$s_1 = 10 + 4(-2)$$
$$s_2 = [10 + 4(-2)] + 4(-1) = 10 + 4(-2 - 1)$$
$$s_3 = [10 + 4(-2 - 1)] + 4(0) = 10 + 4(-2 - 1 + 0)$$
$$s_4 = [10 + 4(-2 - 1 + 0)] + 4(1) = 10 + 4(-2 - 1 + 0 + 1)$$
$$\vdots$$

Thus we conjecture that $s_n = 10 + 4[-2 - 1 + 0 + \cdots + (n - 3)]$ is a solution to the recurrence relation for $n \geq 0$. Using the formula from Exercise 11 in Section 2.5, we can write this conjecture more compactly as

$$10 + 4[-2 - 1 + 0 + \cdots + (n - 3)] = 10 + 4(-2 - 1 + 0) + 4[1 + 2 + \cdots + (n - 3)]$$
$$= 10 - 12 + 4 \left[\frac{(n - 3)(n - 2)}{2} \right]$$
$$= -2 + 2(n - 3)(n - 2)$$
$$= -2 + 2(n^2 - 5n + 6)$$
$$= 2n^2 - 10n + 10.$$

This conjecture can be established by mathematical induction.

21. From the recurrence relation and the initial condition, we have

$$s_0 = 1$$
$$s_1 = -1 + a^1 = a - 1$$
$$s_2 = -(a - 1) + a^2 = a^2 - a + 1$$
$$s_3 = -(a^2 - a + 1) + a^3 = a^3 - a^2 + a - 1$$
$$\vdots$$

Thus we conjecture that $s_n = a^n - a^{n-1} + \cdots + (-1)^n$ is a solution to the recurrence relation for $n \geq 0$. Using the formula from Example 2.51, we can write this conjecture more compactly as

$$a^n - a^{n-1} + \cdots + (-1)^n = (-1)^n (1 - a + a^2 - \cdots + a^n)$$

$$= (-1)^n \left[\frac{(-a)^{n+1} - 1)}{(-a) - 1} \right]$$

$$= (-1)^{n+1} \frac{(-a)^{n+1} - 1}{a + 1}$$

$$= \frac{a^{n+1} + (-1)^n}{a + 1}.$$

This conjecture can be established by mathematical induction.

23. From the recurrence relation and the initial condition, we have

$$s_0 = 3$$

$$s_1 = 1(3) + 1 = 1! \left(3 + \frac{1}{1!} \right)$$

$$s_2 = 2 \cdot 1! \left(3 + \frac{1}{1!} \right) + 1 = 2! \left(3 + \frac{1}{1!} \right) + 1 = 2! \left(3 + \frac{1}{1!} + \frac{1}{2!} \right)$$

$$s_3 = 3 \cdot 2! \left(3 + \frac{1}{1!} + \frac{1}{2!} \right) + 1 = 3! \left(3 + \frac{1}{1!} + \frac{1}{2!} + \frac{1}{3!} \right)$$

$$\vdots$$

Thus we conjecture that

$$s_n = n! \left[3 + \frac{1}{1!} + \frac{1}{2!} + \frac{1}{3!} + \cdots + \frac{1}{n!} \right] = n! \left[2 + \frac{1}{0!} + \frac{1}{1!} + \frac{1}{2!} + \frac{1}{3!} + \cdots + \frac{1}{n!} \right]$$

is a solution to the recurrence relation for $n \geq 0$. This conjecture can be established by mathematical induction.

25. (a) Since the attendance is decreasing by 5 percent per year, each year's attendance is 95 percent of the previous year's. This gives a recurrence relation of $s_n = 0.95 s_{n-1}$ for $n \geq 1$ with an initial condition of $s_0 = 1000$.

(b) Use the method of iteration as in Exercises 11–23 to form the conjecture $s_n = (0.95)^n \cdot 1000$ for $n \geq 0$. This conjecture can be verified by mathematical induction as in Exercises 1–9.

(c) Evaluating the solution for $n = 10$, we find that the predicted enrollment is approximately 599 students.

27. Four toothpicks are needed to make a 1×1 square; thus $s_1 = 4$. For $n \geq 2$, an $n \times n$ square contains an $(n-1) \times (n-1)$ square in its upper left corner. This $(n-1) \times (n-1)$

square contains s_{n-1} toothpicks. At the bottom of the $n \times n$ square, $n+1$ vertical toothpicks and n horizontal toothpicks are still uncounted. Also, At the right of the $n \times n$ square, $n-1$ vertical toothpicks and n horizontal toothpicks are still uncounted. The total number of these uncounted toothpicks is

$$(n+1) + n + (n-1) + n = 4n.$$

Therefore the number of toothpicks required to make an $n \times n$ square is $s_n = s_{n-1} + 4n$. Thus the recurrence relation is $s_n = s_{n-1} + 4n$ for $n \geq 2$, with the initial condition $s_1 = 4$.

Applying the method of iteration, we see that

$$\begin{aligned}
s_1 &= 4 \\
s_2 &= 4 + 4(2) = 4(1) + 4(2) \\
s_3 &= [4(1) + 4(2)] + 4(3) = 4(1) + 4(2) + 4(3) \\
s_4 &= [4(1) + 4(2) + 4(3)] + 4(4) = 4(1) + 4(2) + 4(3) + 4(4) \\
&\;\;\vdots
\end{aligned}$$

Thus we conjecture that $s_n = 4(1) + 4(2) + \cdots + 4n$ is a solution to the recurrence relation for $n \geq 0$. Using the formula from Exercise 11 in Section 2.5, we can write this expression as

$$\begin{aligned}
s_n &= 4(1) + 4(2) + \cdots + 4n \\
&= 4(1 + 2 + \cdots + n) \\
&= 4 \left[\frac{n(n+1)}{2} \right] \\
&= 2n(n+1) \\
&= 2n^2 + 2n.
\end{aligned}$$

This conjecture can be established by mathematical induction.

29. By an *acceptable* sequence, we will mean a sequence of terms in which each term is $-1, 0$, or 1 and no term of 0 follows a term of 1. Note that every sequence of length 1 is acceptable; so $s_1 = 3^1 = 3$. Also, every sequence of length 2 is acceptable except 10; so $s_2 = 3^2 - 1 = 8$. For $n \geq 3$, consider an acceptable sequence of n terms. If the last term of such a sequence is -1 or 1, then the preceding terms can be any acceptable sequence of $n-1$ terms. Thus there are $2 \cdot s_{n-1}$ acceptable sequences of n terms in which the last term is -1 or 1. Now consider an acceptable sequence of n terms in which the last term is 0. Since no term of 0 can follow a term of 1, each of the first $n-1$ terms of this sequence must be -1 or 0. The number of such sequences is 2^{n-1}. Hence s_n satisfies the recurrence relation $s_n = 2s_{n-1} + 2^{n-1}$ for $n \geq 2$ with the initial condition $s_1 = 3$.

To obtain a formula for s_n we note that

$$s_1 = 3 = 2 + 1$$
$$s_2 = 2(2 + 1) + 2^1 = 2^2 + 2^1 + 2^1 = 2^2 + 2(2^1)$$
$$s_3 = 2[2^2 + 2(2^1)] + 2^2 = 2^3 + 2(2^2) + 2^2 = 2^3 + 3(2^2)$$
$$s_4 = 2[2^3 + 3(2^2)] + 2^3 = 2^4 + 3(2^3) + 2^3 = 2^4 + 4(2^3)$$
$$\vdots$$

Thus we conjecture that

$$s_n = 2^n + n2^{n-1} = 2^{n-1}(2 + n).$$

This conjecture can be established by mathematical induction.

31. Examination of the first three cases shows that $r_1 = 2$, $r_2 = 4$, and $r_3 = 7$. For $n \geq 2$, suppose that n lines are drawn so that every pair of lines intersects and no three lines have a point in common. Select $n - 1$ of these lines, which form r_{n-1} regions. Line n intersects each of the previous $n - 1$ lines, and in so doing, it subdivides an existing region into two parts. It also subdivides an additional unbounded region after crossing all of the other $n - 1$ lines. Thus, in addition to the r_{n-1} regions formed by the first $n - 1$ lines, there are n additional regions formed by line n. Hence the recurrence relation and its initial condition are $r_n = r_{n-1} + n$ for $n \geq 1$ and $r_1 = 2$. The method of iteration can be applied as in Example 9.10 to conjecture that

$$\frac{n^2 + n + 2}{2}$$

is the number of regions created by n lines, and this conjecture can be established by mathematical induction.

33. Consider the minimum number of moves m_n necessary to move $2n$ disks from spoke 1 to spoke 3. For $n = 1$, we may move the top disk from spoke 1 to spoke 3 and then move the second disk from spoke 1 on top of it. Thus $m_1 = 2$. For $n = 2$, we must move the top disk from spoke 1 to spoke 2 and move the other disk of the same size on top of it. Then we move one of the largest disks from spoke 1 to spoke 3 and move the other of the same size on top of it. Finally, we move the two disks on spoke 2 to spoke 3. Altogether, this requires 6 moves, and so $m_2 = 6$. For $n \geq 2$, the process is similar to that described in Example 9.2. First, one moves the smallest $2n - 2$ disks from spoke 1 to spoke 2 in m_{n-1} moves. Then the two largest disks are moved from spoke 1 to spoke 3 in 2 moves, and finally the disks on spoke 2 are moved to spoke 3 in m_{n-1} moves. Thus $m_n = 2m_{n-1} + 2$ for $n \geq 1$ with $m_1 = 2$. By a calculation that is similar to that used in obtaining a formula for the ordinary Towers of Hanoi game, we conjecture that

$$m_n = 2^n + 2^{n-1} + \cdots + 2^1 = 2(2^n - 1).$$

193

Note that this number of moves is twice the number of moves in the ordinary Towers of Hanoi game.

9.3 LINEAR DIFFERENCE EQUATIONS WITH CONSTANT COEFFICIENTS

1. Since the coefficient of s_{n-1} is 1, we apply Theorem 9.1 with $a = 1$ and $b = 3$, and $s_0 = 2$ to find s_n. This yields the formula $s_n = 2 + 3n$ for $n \geq 0$.

3. Since the coefficient of s_{n-1} is 4, we apply Theorem 9.1 with $a = 4$ and $b = 0$. Then

$$c = \frac{b}{a-1} = \frac{0}{4-1} = 0.$$

Because $s_0 = 5$, we obtain the formula $s_n = 5(4^n)$ for $n \geq 0$.

5. We apply Theorem 9.1 with $a = -1$ and $b = 6$. In this case

$$c = \frac{b}{a-1} = \frac{6}{-1-1} = -3.$$

Because $s_0 = -4$, we obtain the formula $s_n = 3 - 7(-1)^n$ for $n \geq 0$.

7. We apply Theorem 9.1 with $a = 3$, $b = -8$, and $s_0 = 3$ to obtain the formula $s_n = 4 - 3^n$ for $n \geq 0$.

9. We apply Theorem 9.1 with $a = 1$, $b = -5$, and $s_0 = 100$ to obtain the formula $s_n = 100 - 5n$ for $n \geq 0$.

11. We apply Theorem 9.1 with $a = -2$, $b = -9$, and $s_0 = 7$ to obtain the formula $s_n = 10(-2)^n - 3$ for $n \geq 0$.

13. We use Theorem 9.2 with $a = 1$ and $b = 2$, and so we consider the auxiliary equation $x^2 = x + 2$. Since $x^2 - x - 2 = (x - 2)(x + 1)$, there are distinct roots of 2 and -1. By Theorem 9.2(a), there are constants c_1 and c_2 such that $s_n = c_1(2^n) + c_2(-1)^n$ for $n \geq 0$. Using the initial values of $s_0 = 9$ and $s_1 = 0$, we obtain the system of equations

$$c_1 + c_2 = 9$$
$$2c_1 - c_2 = 0.$$

Solving this system yields $c_1 = 3$ and $c_2 = 6$, and so we obtain the formula $s_n = 3(2^n) + 6(-1)^n$ for $n \geq 0$.

15. We use Theorem 9.2 with $a = 8$ and $b = -16$. Solving the equation $x^2 = 8x - 16$, or equivalently, $0 = x^2 - 8x + 16 = (x - 4)^2$, we see that there is a double root of 4. By

Theorem 9.2(b), we find constants c_1 and c_2 such that $s_n = (c_1 + nc_2)(4)^n$ for $n \geq 0$. Using the initial values $s_0 = 6$ and $s_1 = 20$, we establish the system

$$\begin{aligned} c_1 \quad\quad &= 6 \\ 4c_1 + 4c_2 &= 20. \end{aligned}$$

This system has the solution $c_1 = 6$ and $c_2 = -1$, and so the desired formula is $s_n = (6 - n)(4)^n$ for $n \geq 0$.

17. Since $x^2 - 9 = (x - 3)(x + 3)$, we see that the roots of the auxiliary equation are 3 and -3. By Theorem 9.2(a), we know there are constants c_1 and c_2 such that $s_n = c_1(3^n) + c_2(-3)^n$ for $n \geq 0$. Using the initial conditions $s_0 = 1$ and $s_1 = 9$, we establish the system

$$\begin{aligned} c_1 + \quad c_2 &= 1 \\ 3c_1 - 3c_2 &= 9. \end{aligned}$$

This system has the solution $c_1 = 2$ and $c_2 = -1$, and so $s_n = 2(3^n) - 1(-3)^n$ for $n \geq 0$.

19. Solving the auxiliary equation $x^2 = -4x - 4$, we get a double root of -2. By Theorem 9.2(b), we have $s_n = (c_1 + nc_2)(-2)^n$ for some constants c_1 and c_2. Using the initial conditions $s_0 = -4$ and $s_1 = 2$, we obtain the values $c_1 = -4$ and $c_2 = 3$. Therefore $s_n = (-4 + 3n)(-2)^n$ for $n \geq 0$.

21. Solving the auxiliary equation $x^2 = 10x - 25$, we obtain a double root of 5. So, by Theorem 9.2(b), we see that $s_n = (c_1 + nc_2)(5)^n$ for some constants c_1 and c_2. Using the initial conditions $s_0 = -7$ and $s_1 = -15$, we obtain the values $c_1 = -7$ and $c_2 = 4$. Therefore $s_n = (-7 + 4n)(5)^n$ for $n \geq 0$.

23. Solving the auxiliary equation $x^2 = -5x - 4$, we get roots of -1 and -4. Hence, by Theorem 9.2(a), we have $s_n = c_1(-1)^n + c_2(-4)^n$ for some constants c_1 and c_2. Using the initial conditions $s_0 = 3$ and $s_1 = 15$, we obtain the values $c_1 = 9$ and $c_2 = -6$. Therefore $s_n = 9(-1)^n - 6(-4)^n$ for $n \geq 0$.

25. **(a)** Because 20% of the drug in Mr. Lorenzo's body is eliminated each day, the amount in his body before taking capsule n is 80% of what it was after taking capsule $n - 1$. Capsule n adds another 25 mg of the drug; so $d_n = 0.80d_{n-1} + 25$ for $n \geq 1$ with an initial condition of $d_0 = 0$.

(b) After taking the eighth capsule, the amount of drug in Mr. Lorenzo's body is $d_8 = 104.02848$ mg.

(c) The level to which the drug will eventually accumulate is approximately 125 mg, the limit of $d_n = (0.80)^n(-125) + 125$ as n increases.

27. The value s_n of Michelle's account satisfies $s_n = (1.005)s_{n-1} + 100$ at the start of month n, with an initial value of $s_0 = 1000$. Thus $s_n = 21{,}000(1.005)^n - 20{,}000$ by Theorem 9.1. To determine the value of the account in two years (24 months), we evaluate this expression for $n = 24$, obtaining $\$3670.36$.

29. Let the Johnsons' monthly payment be denoted p. Then the amount owed, a_n, after n payments satisfies $a_n = (1.009)a_{n-1} - p$. Initially, the Johnson family owes $a_0 = \$159,000 - \$32,000 = \$127,000$. Applying Theorem 9.1 to this recurrence relation and initial condition, we obtain

$$a_n = (1.009)^n \left(127,000 - \frac{p}{.009} \right) + \frac{p}{.009}.$$

Because this is a 30-year mortgage, the amount owed after 360 payments will be $0. Solving the preceding equation for p when $n = 360$ and $a_n = 0$, we obtain $p = \$1190.30$. This is the Johnsons' monthly payment.

31. The only way to obtain 1 as a sum of 1s and 2s is with a single 1; so $s_1 = 1$. There are two ways to obtain a sum of 2, namely, $1 + 1$ and 2; so $s_2 = 2$. Suppose that, for $n \geq 3$, we have a sum of 1s and 2s equal to n. If the last summand is 1, then the preceding summands add up to $n - 1$; so there are s_{n-1} ways to obtain a sum of n with the last summand equaling 1. Likewise, if the last summand is 2, then the preceding summands add up to $n - 2$; so there are s_{n-2} ways to obtain a sum of n with the last summand equaling 2. Hence we obtain the recurrence relation $s_n = s_{n-1} + s_{n-2}$ for $n \geq 3$ with the initial conditions $s_1 = 1$, $s_2 = 2$. By a calculation similar to that in Example 9.17, we obtain

$$s_n = \frac{5 + \sqrt{5}}{10} \left(\frac{1 + \sqrt{5}}{2} \right)^n + \frac{5 - \sqrt{5}}{10} \left(\frac{1 - \sqrt{5}}{2} \right)^n.$$

33. For $n = 3$, the three vertices must each have a different color. Thus $c_3 = 4 \cdot 3 \cdot 2 = 24$. For $n = 4$, v_1 and v_3 are either different colors or the same color. If v_1 and v_3 are different colors, then v_4 must be a color other than that used for v_1 or v_3; so there are $4 \cdot 3 \cdot 2 \cdot 2 = 48$ colorings of this type. If v_1 and v_3 are the same color, then v_4 may be any color other than that used for v_1 and v_3; so there are $4 \cdot 3 \cdot 1 \cdot 3 = 36$ colorings of this type. Hence, by the addition principle, $c_4 = 48 + 36 = 84$.

For $n \geq 5$, the situation is similar to the case $n = 4$. Consider a properly colored cycle $v_1, v_2, \ldots, v_{n-1}, v_n, v_1$ of length n. Either v_{n-1} is a different color than v_1 or is the same color as v_1. If v_{n-1} is a different color than v_1, then the color of v_n can be either of the colors not used for v_{n-1} and v_1. Note that, in this case, $v_1, v_2, \ldots, v_{n-1}, v_1$ is a properly colored cycle of length $n - 1$. Thus there are $2 \cdot c_{n-1}$ colorings in which v_1 and v_{n-1} are different colors. If v_{n-1} is the same color as v_1, then v_n can be any color except the color of v_1. In this case, $v_1, v_2, \ldots, v_{n-2}, v_1$ is a properly colored cycle of length $n - 2$; so there are $3 \cdot c_{n-2}$ colorings in which v_1 and v_{n-1} are the same color. Hence c_n satisfies the recurrence relation $c_n = 2c_{n-1} + 3c_{n-2}$ and the initial conditions $c_3 = 24$ and $c_4 = 84$.

This recurrence relation is a second-order homogeneous linear difference equation with constant coefficients; so we may apply Theorem 9.2(a) as in Exercise 13 to obtain a formula for c_n. In this manner, we obtain the formula $c_n = 3^n + 3(-1)^n$, which is valid for $n \geq 3$.

35. We prove the statement by mathematical induction on n, using cases.

Case 1: $a = 1$

When $n = 0$, the formula produces $s_0 + 0 \cdot b = s_0$. Suppose that the formula is correct for a positive integer $n = k$, that is, suppose that $s_k = s_0 + kb$. Consider s_{k+1}. From the recurrence relation, we obtain

$$s_{k+1} = as_k + b = a(s_0 + kb) + b = s_0 + (k+1)b,$$

because $a = 1$. Thus, by the principle of mathematical induction, $s_n = s_0 + nb$ for all nonnegative integers n.

Case 2: $a \neq 1$

When $n = 0$, the formula yields $a^0(s_0 + c) - c = (s_0 + c) - c = s_0$. Suppose that the formula is correct for a positive integer $n = k$, that is, suppose that $s_k = a^k(s_0 + c) - c$, where $c = b/(a-1)$. Consider s_{k+1}. From the recurrence relation, we obtain

$$s_{k+1} = as_k + b = a[a^k(s_0 + c) - c] + b = a^{k+1}(s_0 + c) - ac + b = a^{k+1}(s_0 + c) - c$$

because

$$-ac + b = -a\left(\frac{b}{a-1}\right) + b = \frac{-ab}{a-1} + \frac{b(a-1)}{a-1} = \frac{-b}{a-1} = -c.$$

Therefore $s_n = a^n(s_0 + c) - c$ for all nonnegative integers n by the principle of mathematical induction.

37. Note that, in Theorem 9.2(a), we have $r_1 \neq r_2$, and so $r_2 - r_1 \neq 0$. Substituting the given expressions for c_1 and c_2 into $c_1 r_1^n + c_2 r_2^n$ for $n = 0$ gives

$$c_1 r_1^0 + c_2 r_2^0 = c_1 + c_2 = \frac{r_2 s_0 - s_1}{r_2 - r_1} + \frac{s_1 - r_1 s_0}{r_2 - r_1} = \frac{(r_2 - r_1)s_0}{r_2 - r_1} = s_0.$$

Substituting the given expressions for c_1 and c_2 into $c_1 r_1^n + c_2 r_2^n$ for $n = 1$ gives

$$c_1 r_1^1 + c_2 r_2^1 = \frac{r_2 s_0 - s_1}{r_2 - r_1}(r_1) + \frac{s_1 - r_1 s_0}{r_2 - r_1}(r_2) = \frac{s_1(r_2 - r_1)}{r_2 - r_1} = s_1.$$

9.4 ANALYZING THE EFFICIENCY OF ALGORITHMS WITH RECURRENCE RELATIONS

1. Because the floor of a number x is the greatest integer that is less than or equal to x, we see that $\lfloor 243 \rfloor = 243$.

3. Since $4 \leq \frac{33}{7} < 5$, then, as in Exercise 1, $\left\lfloor \frac{33}{7} \right\rfloor = 4$.

5. Because $0 \le 0.871 < 1$, we have $\lfloor 0.871 \rfloor = 0$.

7. Because $\dfrac{-(-34+2)}{2} = \dfrac{-(-32)}{2} = \dfrac{32}{2} = 16$, we have $\left\lfloor \dfrac{-(-34+2)}{2} \right\rfloor = 16$.

9. We apply the binary search algorithm as in Example 9.22 to search for $t = 83$.

b	e	m	a_m	Is $a_m = t$?
1	100	$\left\lfloor \dfrac{1+100}{2} \right\rfloor = 50$	50	no; $a_m < t$
51	100	$\left\lfloor \dfrac{51+100}{2} \right\rfloor = 75$	75	no; $a_m < t$
76	100	$\left\lfloor \dfrac{76+100}{2} \right\rfloor = 88$	88	no; $a_m > t$
76	87	$\left\lfloor \dfrac{76+87}{2} \right\rfloor = 81$	81	no; $a_m < t$
82	87	$\left\lfloor \dfrac{82+87}{2} \right\rfloor = 84$	84	no; $a_m > t$
82	83	$\left\lfloor \dfrac{82+83}{2} \right\rfloor = 82$	82	no; $a_m < t$
83	83	$\left\lfloor \dfrac{83+83}{2} \right\rfloor = 83$	83	yes

11. We apply the binary search algorithm as in Example 9.22 to search for $t = 400$.

b	e	m	a_m	Is $a_m = t$?
1	300	$\left\lfloor \dfrac{1 + 300}{2} \right\rfloor = 150$	450	no; $a_m > t$
1	149	$\left\lfloor \dfrac{1 + 149}{2} \right\rfloor = 75$	225	no; $a_m < t$
76	149	$\left\lfloor \dfrac{76 + 149}{2} \right\rfloor = 112$	336	no; $a_m < t$
113	149	$\left\lfloor \dfrac{113 + 149}{2} \right\rfloor = 131$	393	no; $a_m < t$
132	149	$\left\lfloor \dfrac{132 + 149}{2} \right\rfloor = 140$	420	no; $a_m > t$
132	139	$\left\lfloor \dfrac{132 + 139}{2} \right\rfloor = 135$	405	no; $a_m > t$
132	134	$\left\lfloor \dfrac{132 + 134}{2} \right\rfloor = 133$	399	no; $a_m < t$
133	134	$\left\lfloor \dfrac{132 + 134}{2} \right\rfloor = 133$	402	no; $a_m > t$
134	133			

Since $b > e$, the target t is not in the list.

13. Using the merging and merge sort algorithms as in Examples 9.23 and 9.24, we obtain

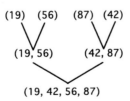

15. Using the merging and merge sort algorithms as in Examples 9.23 and 9.24, we obtain

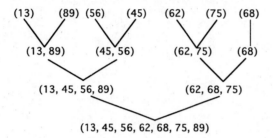

17. Using the merging and merge sort algorithms as in Examples 9.23 and 9.24, we obtain

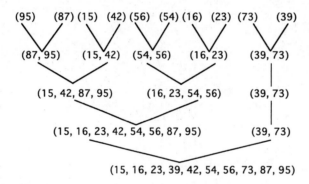

19. For any positive integer n, we have $n \log_2 n < \frac{n^2}{2}$ if and only if $\log_2 n < \frac{n}{2}$. Comparing the graphs of the functions $f(x) = \log_2 x$ and $g(x) = \frac{1}{2}x$, we see that $f(x) < g(x)$ for $4 < x$. Thus the second algorithm is better for all positive integers greater than 4.

21. Let x be a real number. Then there exists a unique integer k such that $k \le x < k+1$. Now $\lfloor x \rfloor = k$. So $\lfloor x \rfloor \le x < \lfloor x \rfloor + 1$.

23. Let x and y be real numbers. There exist unique integers m and n such that

$$m \le x < m+1 \quad \text{and} \quad n \le y < n+1.$$

Then $\lfloor x \rfloor = m$ and $\lfloor y \rfloor = n$. Therefore

$$m + n \le x + y < (m+1) + (n+1) = m + n + 2.$$

Since $\lfloor x + y \rfloor$ is an integer and $x + y < m + n + 2$, we must have

$$\lfloor x + y \rfloor \le m + n + 1 = \lfloor x \rfloor + \lfloor y \rfloor + 1.$$

Equality does not hold when $x = 3$ and $y = -4.5$.

25. In the algorithm for evaluating x^n, the only elementary operation performed when $n = 1$ is the comparison of k to n; hence $e_1 = 1$. Also, there are three elementary operations performed during each complete iteration of step 2 (namely, one comparison of k to n, one multiplication of P and x, and one addition of k and 1). Because step 2 is completed $n - 1$ times in the evaluation of x^n, we obtain the recurrence relation $e_n = e_{n-1} + 3$ for $n \geq 2$ with the initial condition $e_1 = 1$.

27. In Horner's polynomial evaluation algorithm, the only elementary operation performed when $n = 0$ is the comparison of k to n; so $e_0 = 1$. Also, there are five elementary operations performed during each complete iteration of step 2 (namely, one comparison of k to n, one multiplication of x and S, one subtraction of k from n, one addition of xS and a_{n-k}, and one addition of k and 1). Because step 2 is completed n times in evaluating a polynomial of degree n but only the comparison of k to n is performed when $k = n + 1$, we obtain the recurrence relation $e_n = e_{n-1} + 5$ for $n \geq 2$ with the initial condition $e_0 = 1$.

29. Applying Theorem 9.1 with $a = 1$ gives $e_n = 1 + 3(n - 1) = 3n - 2$ for $n \geq 1$.

31. Applying Theorem 9.1 with $a = 1$ gives $e_n = 5n + 1$ for $n \geq 0$.

33. For all positive integers n and k, where $1 \leq k \leq n$, we have $(n - k)(k - 1) \geq 0$ because each factor is nonnegative. Hence $nk - k^2 - n + k \geq 0$. This is equivalent to $k(n + 1 - k) \geq n$.

35. For every positive integer n, we have $(n - 1)^2 \geq 0$. Hence $n^2 - 2n + 1 \geq 0$. It follows that $n^2 + 2n + 1 \geq 4n$, or $(n + 1)^2 \geq 4n$. Taking the positive square root of each side of the inequality yields $n + 1 \geq 2\sqrt{n}$, that is,

$$\frac{n + 1}{2} \geq \sqrt{n}.$$

37. Let $n = 2^k$ for some nonnegative integer k. The proof is by induction on k. Define $x_n = 1 + n(\log_2 n - 1)$. For $k = 0$, we have $n = 1$, and

$$x_n = x_1 = 1 + 1(\log_2 1 - 1) = 1 + 1(0 - 1) = 0 = s_1.$$

Assume that x_n is a solution to the recurrence relation $s_n = 2s_{\lfloor n/2 \rfloor} + (n - 1)$ for some nonnegative integer k. Then

$$s_{k+1} = 2x_{\lfloor 2^{k+1}/2 \rfloor} + (2^{k+1} - 1)$$
$$= 2s_k + (2^{k+1} - 1)$$
$$= 2x_{2^k} + (2^{k+1} - 1)$$
$$= 2[1 + 2^k(\log_2 2^k - 1)] + (2^{k+1} - 1)$$
$$= 2 + 2^{k+1}(\log_2 2^k - 1) + (2^{k+1} - 1)$$
$$= 1 + 2^{k+1}(k - 1) + 2^{k+1}$$
$$= 1 + 2^{k+1}k$$
$$= 1 + 2^{k+1}[(k+1) - 1]$$
$$= 1 + 2^{k+1}(\log_2 2^{k+1} - 1) = x_{k+1}.$$

Hence, by the principle of mathematical induction, x_n is a solution to the recurrence relation $s_n = 2s_{\lfloor n/2 \rfloor} + (n - 1)$ for $n = 2^k \geq 2$ with the initial condition $s_1 = 0$.

39. The proof is by mathematical induction on n. For $n = 1$,

$$k + c\lfloor \log_2 1 \rfloor = k = s_1,$$

and so the given expression is a solution to the recurrence relation. Assume that for all positive integers less than or equal to n, the given expression is a solution to the recurrence relation, that is, $s_r = k + c\lfloor \log_2 r \rfloor$ for $r \leq n$. Consider s_{n+1}. If n is even, then $n + 1$ is odd and $n = 2m$ for some positive integer m. Thus

$$s_{n+1} = s_{2m+1} = s_{\lceil m+.5 \rceil} + c$$
$$= (k + c\lfloor \log_2 (m + .5) \rfloor) + c$$
$$= k + c(\lfloor \log_2 (m + .5) \rfloor + 1)$$
$$= k + c(\lfloor \log_2 (m + .5) \rfloor + \log_2 2)$$
$$= k + c\lfloor \log_2 2(m + .5) \rfloor)$$
$$= k + c\lfloor \log_2 (2m + 1) \rfloor$$
$$= k + c\lfloor \log_2 (n + 1) \rfloor.$$

Therefore, when n is even, the given expression is a solution to the recurrence relation for $n + 1$.

If n is odd, then $n + 1$ is even and $n + 1 = 2m$ for some positive integer m. So

$$
\begin{aligned}
s_{n+1} = s_{2m} \\
&= (k + c\lfloor \log_2 m \rfloor) + c \\
&= k + c(\lfloor \log_2 (m + .5) \rfloor + 1) \\
&= k + c\lfloor \log_2 m + 1 \rfloor \\
&= k + c\lfloor \log_2 m + \log_2 2 \rfloor \\
&= k + c\lfloor \log_2 2m \rfloor \\
&= k + c\lfloor \log_2 (n + 1) \rfloor.
\end{aligned}
$$

Therefore, when n is odd, the given expression is a solution to the recurrence relation for $n + 1$. Hence, by the strong principle of mathematical induction, the given expression is a solution to the recurrence relation for all positive integers n.

9.5   COUNTING WITH GENERATING FUNCTIONS

1. We have

$$
\begin{aligned}
A + B &= (1 + 1) + (1 + 2)x + (1 + 0)x^2 + (0 + 0)x^3 + (0 + 4)x^4 + (0 + 1)x^5 \\
&= 2 + 3x + x^2 + 4x^4 + x^5.
\end{aligned}
$$

3. We have

$$
\begin{aligned}
AB &= (1 + x + x^2)B \\
&= B + xB + x^2 B \\
&= (1 + 2x + 4x^4 + x^5) + (x + 2x^2 + 4x^5 + x^6) + (x^2 + 2x^3 + 4x^6 + x^7) \\
&= 1 + 3x + 3x^2 + 2x^3 + 4x^4 + 5x^5 + 5x^6 + x^7.
\end{aligned}
$$

5. We have

$$
\begin{aligned}
B + D &= (1 + 1) + (2 + 1)x + (0 + 1)x^2 + (0 + 1)x^3 + (4 + 1)x^4 + (1 + 1)x^5 \\
&\quad + (0 + 1)x^6 + (0 + 1)x^7 + \cdots \\
&= 2 + 3x + x^2 + x^3 + 5x^4 + 2x^5 + x^6 + x^7 + \cdots.
\end{aligned}
$$

7. We have

$$AD = (1 + x + x^2)D$$
$$= D + xD + x^2 D$$
$$= (1 + x + x^2 + x^3 + x^4 + x^5 + \cdots) + (x + x^2 + x^3 + x^4 + x^5 + \cdots)$$
$$+ (x^2 + x^3 + x^4 + x^5 + \cdots)$$
$$= 1 + 2x + 3x^2 + 3x^3 + 3x^4 + 3x^5 + 3x^6 + 3x^7 + \cdots.$$

9. We have

$$EC = (1 + x^3 + x^6 + \cdots)C$$
$$= C + x^3 C + x^6 C + \cdots$$
$$= (1 - x^2 + x^4) + (x^3 - x^5 + x^7) + (x^6 - x^8 + x^{10}) + \cdots$$
$$= 1 - x^2 + x^3 + x^4 - x^5 + x^6 + x^7 - \cdots.$$

11. We have

$$DE = D(1 + x^3 + x^6 + \cdots)$$
$$= C + Dx^3 + Dx^6 + \cdots$$
$$= (1 + x + x^2 + x^3 + x^4 + x^5 + x^6 + x^7 + \cdots) + (x^3 + x^4 + x^5 + x^6 + x^7 + \cdots)$$
$$+ (x^6 + x^7 + \cdots)$$
$$= 1 + x + x^2 + 2x^3 + 2x^4 + 2x^5 + 3x^6 + 3x^7 + \cdots.$$

13. In the product $(1 + x + x^2 + x^3)(1 + x + x^2 + x^3 + x^4 + x^5)$, which equals

$$1 + 2x + 3x^2 + 4x^3 + 4x^4 + 4x^5 + 3x^6 + 2x^7 + x^8,$$

the first factor represents the number of Cokes selected and the second represents the number of Pepsis selected.

15. In the product $(1 + x + x^2 + x^3)(1 + x + x^2 + x^3 + x^4)(1 + x + x^2)$, which equals

$$1 + 3x + 6x^2 + 9x^3 + 11x^4 + 11x^5 + 9x^6 + \cdots,$$

the first factor represents the number of licorice pieces, the second the number of strawberry pieces, and the third the number of lemon pieces.

17. In the product $(1 + x + x^2 + x^3 + x^4)(1 + x + x^2 + x^3)(1 + x^2)(1 + x^3)$, which equals

$$1 + 2x + 4x^2 + 7x^3 + 9x^4 + 11x^5 + 12x^6 + \cdots,$$

the first factor represents the number of wings, the second the number of breasts, the third the number of packages of two drumsticks, and the fourth the number of packages of three drumsticks selected.

19. In the product $(1 + x + x^2 + x^3)(1 + x + x^2 + x^3 + x^4 + x^5 + \cdots)$, which equals

$$1 + 2x + 3x^2 + 4x^3 + 4x^4 + 4x^5 + 4x^6 + \cdots,$$

the first factor represents the number of glasses of milk and the second factor represents the number of glasses of water.

21. In the product: $(x^4 + x^5 + x^6 + \cdots)(x^2 + x^3 + x^4 + \cdots)$, which equals

$$x^6 + 2x^7 + 3x^8 + 4x^9 + \cdots,$$

the first factor represents the number of oak leaves selected and the second represents the number of maple leaves selected.

23. In the product:

$$(1 + x)(1 + x)(1 + x)(1 + x)(1 + x)(1 + x)(1 + x)(1 + x + x^2 + x^3 + x^4 + x^5)$$
$$= 1 + 8x + 29x^2 + 64x^3 + 99x^4 + 120x^5 + 126x^6 + \cdots,$$

the $(1 + x)$ factors represent the number of each mathematics book chosen and the other factor represents the number of copies of *Peyton Place*.

25. In the product

$$(1 + x + x^2 + x^3 + \cdots)(1 + x^3 + x^6 + x^9 + \cdots)(1 + x^4 + x^8 + x^{12} + \cdots)$$
$$= 1 + x + x^2 + 2x^3 + 3x^4 + 3x^5 + 4x^6 + \cdots,$$

the first factor represents the number of pounds of bluegills, the second the number of pounds of catfish, the third the number of pounds of bass.

27. Note that when $(1 + x + x^2 + x^3 + \cdots)^2$ is multiplied out, the x^r terms are $1x^r, x x^{r-1}$, $x^2 x^{r-2}$, ..., $x^r 1$, that is, $x^i x^{r-i}$, where $i = 0, 1, \ldots, r$. Since there are $r + 1$ of these terms, the coefficient of x^r is $a_r = r + 1$.

29. We have

$$F = (1 + x + x^2 + x^3 + \cdots)(1 + x) = (1 + x + x^2 + x^3 + \cdots) + (x + x^2 + x^3 + \cdots).$$

Clearly $a_0 = 1$ and $a_r = 2$ for $r > 0$.

31. We have

$$F = (1 - x + x^2 - x^3 + \cdots)(1 + x) = (1 - x + x^2 - x^3 + \cdots) + (1 - x + x^2 - x^3 + \cdots)x$$
$$= (1 - x + x^2 - x^3 + \cdots) + (x - x^2 + x^3 - x^4 + \cdots)$$
$$= 1 + 0x + 0x^2 + 0x^3 + \cdots.$$

Clearly $a_0 = 1$ and $a_r = 0$ for $r > 0$.

33. Since the terms p and q are prime numbers, they are elements of the set $\{2, 3, 5, 7, \ldots\}$. Thus there are two factors of $(x^2 + x^3 + x^5 + x^7 + \cdots)$ in the generating function, which is

$$(x^2 + x^3 + x^5 + x^7 + \cdots)(x^2 + x^3 + x^5 + x^7 + \cdots) = x^4 + 2x^5 + x^6 + 2x^7 + 2x^8 + 2x^9 + 3x^{10} + \cdots.$$

35. Since the four addends are squared numbers, we wish to select addends from the set $\{0, 1, 4, 9, \ldots\}$. There is a factor of $(1 + x + x^4 + x^9 + x^{16} + \cdots)$ in the generating function for each addend. It follows that

$$F = (1 + x + x^4 + x^9 + x^{16} + \cdots)^4$$
$$= 1 + 4x + 6x^2 + 4x^3 + 5x^4 + 12x^5 + 12x^6 + 4x^7 + 6x^8 + 16x^9 + 18x^{10} + \cdots.$$

9.6 THE ALGEBRA OF GENERATING FUNCTIONS

1. From equation (9.6), we have

$$(1 - 3x)^{-1} = 1 + (3x) + (3x)^2 + (3x)^3 + \cdots = 1 + 3x + 9x^2 + 27x^3 + \cdots.$$

3. We have
$$1 + 2x + 4x^2 + 8x^3 + \cdots = 1 + (2x) + (2x)^2 + (2x)^3 + \cdots,$$

and so, by equation (9.6), the inverse is $1 - 2x$.

5. By equation (9.6),

$$(1 + x^2)^{-1} = (1 - (-x^2))^{-1} = 1 + (-x^2) + (-x^2)^2 + (-x^2)^3 + \cdots$$
$$= 1 - x^2 + x^4 - x^6 + x^8 - x^{10} + \cdots.$$

7. We have

$$(1 - x - x^2)^{-1} = (1 - (x + x^2))^{-1} = 1 + (x + x^2) + (x + x^2)^2 + (x + x^2)^3 + \cdots$$
$$= 1 + (x + x^2) + (x^2 + 2x^3 + x^4) + (x^3 + 3x^4 + 3x^5 + x^6) + \cdots$$
$$= 1 + x + 2x^2 + 3x^3 + 5x^4 + 8x^5 + \cdots.$$

9. We have

$$(2 + 6x)^{-1} = 2^{-1}(1 - (-3x))^{-1} = 2^{-1}(1 + (-3x) + (-3x)^2 + (-3x)^3 + \cdots)$$
$$= \frac{1}{2} - \frac{3}{2} + \frac{9}{2} - \frac{27}{2} + \cdots.$$

11. We have

$$S = 1 + (2s_0 + 1)x + (2s_1 + 1)x^2 + (2s_2 + 1)x^3 + \cdots$$
$$= 2x(s_0 + s_1 x + s_2 x^2 + s_3 x^3 + \cdots) + 1 + x + x^2 + x^3 + \cdots$$
$$= 2xS + (1 - x)^{-1},$$

so that $S(1 - 2x) = (1 - x)^{-1}$. Therefore $S = (1 - x)^{-1}(1 - 2x)^{-1}$.

13. We have

$$S = 1 + x + (2s_1 - s_0)x^2 + (2s_2 - s_1)x^3 + \cdots$$
$$= 1 + x + 2x(s_1 x + s_2 x^2 + \cdots) - x^2(s_0 + s_1 x + s_2 x^2 + \cdots)$$
$$= 1 + x + 2x(S - 1) - x^2 S = 1 + x - 2x + S(2x - x^2).$$

Thus $S(1 - 2x + x^2) = 1 - x$, and so

$$S = (1 - x)(1 - 2x + x^2)^{-1} = (1 - x)/(1 - x)^2 = (1 - x)^{-1}.$$

15. We have

$$S = -1 + 0x + (-s_1 + 2s_0)x^2 + (-s_2 + 2s_1)x^3 + \cdots$$
$$= -1 - x(s_1 x + s_2 x^2 + \cdots) + 2x^2(s_0 + s_1 x + s_2 x^2 + \cdots)$$
$$= -1 - x(S + 1) + 2x^2 S$$
$$= -1 - x + S(-x + 2x^2).$$

Thus $S(1 + x - 2x^2) = -1 - x$, and so

$$S = -(1 + x)(1 + x - 2x^2)^{-1} = -(1 + x)(1 - x)^{-1}(1 + 2x)^{-1}.$$

17. We have $S = -2 + x + (s_1 + 3s_0 + 2)x^2 + (s_2 + 3s_1 + 2)x^3 + \cdots$. Thus

$$S = -2 + x + x(S + 2) + 3x^2(S) + 2x^2(1 + x + x^2 + \cdots)$$
$$= -2 + 3x + (x + 3x^2)S + \frac{2x^2}{1 - x}.$$

Therefore

$$S(1 - x - 3x^2) = -2 + 3x + \frac{2x^2}{1 - x} = \frac{-2 + 5x - x^2}{1 - x},$$

and so

$$S = \frac{-2 + 5x - x^2}{(1 - x - 3x^2)(1 - x)}.$$

19. We have $S = 2 - x + x^2 + (s_2 - 3s_1 + s_0)x^3 + (s_3 - 3s_2 + s_1)x^4 + \cdots$. Thus

$$S = 2 - x + x^2 + x(S - 2 + x) - 3x^2(S - 2) + x^3 S = 2 - 3x + 8x^2 + (x - 3x^2 + x^3)S,$$

and so $(1 - x + 3x^2 - x^3)S = 2 - 3x + 8x^2$, and

$$S = \frac{2 - 3x + 8x^2}{1 - x + 3x^2 - x^3}.$$

21. We have

$$\frac{x}{(1-x)(1+2x)} = \frac{(1+2x)a}{(1+2x)(1-x)} + \frac{(1-x)b}{(1-x)(1+2x)} = \frac{a + b + (2a - b)x}{(1-x)(1+2x)}.$$

Setting $a + b = 0$ and $2a - b = 1$, we find $a = 1/3$ and $b = -1/3$.

23. We have

$$\frac{1 + 3x}{(1+2x)(1-x)} = \frac{(1-x)a}{(1-x)(1+2x)} + \frac{(1+2x)b}{(1+2x)(1-x)} = \frac{a + b + (-a + 2b)x}{(1+2x)(1-x)}.$$

Setting $a + b = 1$ and $-a + 2b = 3$, we find $a = -\frac{1}{3}$ and $b = \frac{4}{3}$.

25. We have

$$\frac{1 + x}{(1+2x)^2} = \frac{(1+2x)a}{(1+2x)(1+2x)} + \frac{b}{(1+2x)^2} = \frac{a + b + 2ax}{(1+2x)^2}.$$

Setting $a + b = 1$ and $2a = 1$, we find $a = \frac{1}{2}$ and $b = \frac{1}{2}$.

27. Expanding, $S = (1 + 2x + 4x^2 + \cdots) + (1 - x + x^2 - x^3 + \cdots)$. Examining the corresponding s_n, we see that $s_n = 2^n + (-1)^n$.

29. We have

$$S = -1(1 - 2x)^{-1} + 4(1 - (-5x))^{-1}$$
$$= -1(1 + 2x + (2x)^2 + \cdots) + 4(1 + (-5x) + (-5x)^2 + \cdots).$$

Therefore $s_n = -2^n + 4(-5)^n$.

31. We have $S = 2(1 + 3x^2 + 9x^4 + \cdots)$, and so $s_n = 2 \cdot 3^{n/2}$ for even n and $s_n = 0$ for odd n.

33. We have $b_3 = -2b_2 - 3b_1 - 4b_0 = -2(1) - 3(-2) - 4(1) = 0$. Suppose that $b_i = 0$ for every positive integer i less than or equal to k, where $k \geq 3$. Consider b_{k+1}. By the recurrence relation,

$$b_{k+1} = -2(0) - 3(0) - \cdots - (k-1)b_2 - kb_1 - (k+1)b_0$$
$$= -(k-1)(1) - k(-2) - (k+1)(1)$$
$$= -k + 1 + 2k - k - 1$$
$$= 0.$$

Therefore $b_{k+1} = 0$ when $b_i = 0$ for each i such that $3 \leq i \leq k$. Thus, by the strong principle of mathematical induction, $b_n = 0$ for every integer $n \geq 3$.

35. If $k_1 = s_0 + \frac{b}{a-1}$ and $k_2 = \frac{-b}{a-1}$, then

$$\frac{k_1}{1-ax} + \frac{k_2}{1-x} = \frac{k_1(1-x) + k_2(1-ax)}{(1-ax)(1-x)} = \frac{k_1 + k_2 + (-k_1 - ak_2)x}{(1-ax)(1-x)}$$

$$= \frac{s_0 + (-s_0 + \frac{-b+ab}{a-1})}{(1-ax)(1-x)} = \frac{s_0 + (-s_0 + b)x}{(1-ax)(1-x)}.$$

37. By equating the corresponding coefficients of $x^2 - ax - b = (x - r_1)(x - r_2)$, we obtain

$$r_1 + r_2 = a \qquad \text{and} \qquad r_1 r_2 = -b.$$

Now $(1 - r_1 x)(1 - r_2 x) = 1 - (r_1 + r_2)x + r_1 r_2 x^2$. Substituting from the preceding equations yields $1 - ax - bx^2$.

39. Set $c_1 = \dfrac{s_1 + (a + r_1)s_0}{r_1 - r_2}$ and $c_2 = \dfrac{-s_1 - (a + r_2)s_0}{r_1 - r_2}$. Then

$$(1 - r_2 x)c_1 + (1 - r_1 x)c_2 = \frac{(1 - r_2 x)(s_1 + (a + r_1)s_0) + (1 - r_1 x)(-s_1 - (a + r_2)s_0)}{r_1 - r_2}$$

$$= \frac{s_1 + (a + r_1)s_0 - s_1 - (a + r_2)s_0 - r_2(s_1 + (a + r_1)s_0)x + r_1(s_1 + (a + r_2)s_0)x}{r_1 - r_2}$$

$$= \frac{(r_1 - r_2)s_0 + [(r_1 - r_2)s_1 + (r_1 - r_2)as_0]x}{r_1 - r_2} = s_0 + (s_1 + as_0)x.$$

Dividing both sides by $(1 - r_1 x)(1 - r_2 x)$ gives the desired equation.

41. Set $k_1 = \dfrac{-s_1 - as_0}{r}$ and $k_2 = \dfrac{s_1 + (a + r)s_0}{r}$. Then

$$(1 - rx)k_1 + k_2 = \frac{(1 - rx)(-s_1 - as_0) + s_1 + (a + r)s_0}{r}$$

$$= \frac{-s_1 - as_0 + s_1 + (a + r)s_0 + r(s_1 + as_0)x}{r}$$

$$= \frac{rs_0 + r(s_1 + as_0)x}{r} = s_0 + (s_1 + as_0)x.$$

Dividing both sides by $(1 - rx)^2$ gives the desired equation.

43. From Exercise 42, $s_n = k_1 r^n + nk_2 r^n + k_2 r^n$. Collecting terms, we obtain

$$s_n = (k_1 + k_2)r^n + nk_2 r^n,$$

where k_1 and k_2 are as in Exercise 41. Thus we may take $c_1 = k_1 + k_2$ and $c_2 = k_2$.

SUPPLEMENTARY EXERCISES

1. The initial condition is
$$s_0 = 2.$$
By iterating the recurrence relation, we obtain
$$s_1 = 3s_0 + 1^2 = 3(2) + 1 = 7$$
$$s_2 = 3s_1 + 2^2 = 3(7) + 4 = 25$$
$$s_3 = 3s_2 + 3^2 = 3(25) + 9 = 84$$
$$s_4 = 3s_3 + 4^2 = 3(84) + 16 = 268$$
$$s_5 = 3s_4 + 5^2 = 3(268) + 25 = 829.$$

3. Proceeding as in Exercise 1, we obtain the values $s_0 = 1$, $s_1 = 2$, $s_2 = 8$, $s_3 = 48$, $s_4 = 384$, and $s_5 = 3840$.

5. Proceeding as in Exercise 1, we obtain the values $s_0 = 1$, $s_1 = 1$, $s_2 = 1$, $s_3 = 2$, $s_4 = 7$, and $s_5 = 33$.

7. Since the cost of living adjustment is 4 percent of the present year's salary, we have to multiply the present salary by 1.04 and then add the yearly increment of $500 to get the next year's salary. This results in the recurrence relation $s_n = 1.04s_{n-1} + 500$ for $n \geq 2$ with an initial value of $s_1 = 16,000$.

9. We will call a codeword of dots and dashes having no two consecutive dashes *acceptable*. Every sequence of 1 symbol is acceptable; so $c_1 = 2$. In addition, every sequence of 2 symbols except two dashes is acceptable; so $c_2 = 2^2 - 1 = 3$. Consider an acceptable codeword with n symbols. If its last symbol is a dot, then the beginning symbols may be any acceptable sequence of $n - 1$ symbols. Hence there are c_{n-1} codewords ending in a dot. If the last symbol is a dash, then the next-to-last symbol must be a dot, and the beginning symbols may be any acceptable sequence of $n - 2$ symbols. Thus there are c_{n-2} codewords ending with a dot followed by a dash. Therefore the recurrence relation is $c_n = c_{n-1} + c_{n-2}$ for $n \geq 3$ with initial conditions of $c_1 = 2$ and $c_2 = 3$.

11. This relationship involves an incremental loss at each stage that is inversely proportional to the value of n^2. Thus the incremental loss is $\frac{k}{n^2}$, where k is the constant of proportionality. Hence the recurrence relation has the form $e_n = e_{n-1} - \frac{k}{n^2}$ for $n \geq 1$.

13. Because $2(0)^2 + 2(0) = 0 = s_0$, the given expression is a solution for $n = 0$. Suppose that the given expression is a solution for $n = k \geq 1$, that is, $s_k = 2k^2 + 2k$. Applying the recurrence relation, we have
$$s_{k+1} = 2k^2 + 2k + 4(k + 1) = 2k^2 + 2k + 2 + 2k + 2 = 2(k + 1)^2 + 2(k + 1).$$
Hence the given expression is a solution for $n = k + 1$ and so, by the the principle of mathematical induction, for all nonnegative integers.

15. The given expression is a solution for $n = 0$ because $1! - 1 = 0 = s_0$. Suppose that the given expression is a solution for $n = k \geq 1$, that is, $s_k = (k+1)! - 1$. Applying the recurrence relation, we have

$$s_{k+1} = [(k+1)! - 1] + (k+1)(k+1)! = (k+1)![k+1+1] - 1 = (k+2)! - 1.$$

Hence the given expression is a solution for $n = k + 1$ and so, by the principle of mathematical induction, for all nonnegative integers.

17. The given recurrence relation is a first-order linear difference equation with constant coefficients. Applying Theorem 9.1 with $a = 3$, $b = -12$, and $s_0 = 5$, we see that its general term satisfies

$$s_n = 3^n \left(5 + \frac{-12}{3-1}\right) - \frac{-12}{3-1} = 3^n(-1) + 6$$

for $n \geq 0$.

19. The given recurrence relation is a second-order homogeneous linear difference equation with constant coefficients with $x^2 = 4x - 4$ as its auxiliary equation. Because

$$0 = x^2 - 4x + 4 = (x-2)^2,$$

the solution of the recurrence relation has the form $s_n = (c_1 + nc_2)2^n$ for some constants c_1 and c_2. Substituting $n = 0$ and $n = 1$ into the preceding equation and using the initial conditions $s_0 = 4$ and $s_1 = 6$, we obtain the following system of linear equations:

$$c_1 \qquad = 4$$
$$2c_1 + 2c_2 = 6.$$

The unique solution of this system is $c_1 = 4$ and $c_2 = -1$. Hence the general term of a solution is $s_n = (4-n)2^n$ for $n \geq 0$.

21. Each term in the Lucas sequence after the first two is the sum of the two preceding terms. Thus

$$L_1 = 3,$$
$$L_2 = 4,$$
$$L_3 = L_1 + L_2 = 3 + 4 = 7,$$
$$L_4 = L_2 + L_3 = 4 + 7 = 11,$$
$$L_5 = L_3 + L_4 = 7 + 11 = 18,$$
$$L_6 = L_4 + L_5 = 11 + 18 = 29,$$
$$L_7 = L_5 + L_6 = 18 + 29 = 47,$$
$$L_8 = L_6 + L_7 = 29 + 47 = 76,$$
$$L_9 = L_7 + L_8 = 47 + 76 = 123,$$
$$L_{10} = L_8 + L_9 = 76 + 123 = 199.$$

23. Note that $L_1 = p$, $L_2 = q$, and $L_3 = p + q = pF_0 + qF_1$. Suppose that for some positive integer $k \geq 3$, the desired relationship holds for all positive integers $n \leq k$. By definition of a Lucas sequence, we have

$$L_{k+1} = L_k + L_{k-1} = (qF_{k-1} + pF_{k-2}) + (qF_{k-2} + pF_{k-3}) = qF_k + pF_{k-1}.$$

Hence, by the strong principle of mathematical induction, the statement is true for all positive integers $n \geq 3$.

25. The expression r^n satisfies (9.7) if and only if $r^n = a_1 r^{n-1} + a_2 r^{n-2} + \cdots + a_k r^{n-k}$ for all $n \geq k$. But whether $r = 0$ or not, this is equivalent to $r^k = a_1 r^{k-1} + a_2 r^{k-2} + \cdots + a_k$, so that r is a root of the auxiliary equation.

27. The roots of the auxiliary equation $x^3 - 3x^2 - 10x + 24 = 0$ are -3, 2, and 4. Using the results of Exercises 25 and 26, we look for a solution of the form

$$s_n = c_1(-3)^n + c_2(2)^n + c_3(4)^n$$

for some constants c_1, c_2, and c_3. Using the initial conditions, we see that c_1, c_2, and c_3 must satisfy the system

$$\begin{aligned} c_1 + c_2 + c_3 &= -4 \\ -3c_1 + 2c_2 + 4c_3 &= -9 \\ 9c_1 + 4c_2 + 16c_3 &= 13. \end{aligned}$$

The unique solution of this system is $c_1 = 1$, $c_2 = -7$, and $c_3 = 2$. Thus the desired solution is $s_n = (-3)^n - 7(2)^n + 2(4)^n$.

29. We proceed as in Exercise 27 to find a solution of the form

$$s_n = (c_1 + c_2 n)(-1)^n + c_3(2)^n,$$

where -1 and 2 are double and single roots of the auxiliary equation $x^3 - 3x - 2 = 0$, respectively. Using the initial conditions, we obtain the system

$$\begin{aligned} c_1 + c_3 &= 4 \\ -c_1 - c_2 + 2c_3 &= 4 \\ c_1 + 2c_2 + 4c_3 &= -3. \end{aligned}$$

The unique solution of this system is $c_1 = 3$, $c_2 = -5$, and $c_3 = 1$. Thus the desired solution is $s_n = (3 - 5n)(-1)^n + 2^n$.

31. (a) Suppose for the moment that there are values of a and b such that $an + b$ satisfies $s_n = s_{n-1} + 6s_{n-2} + 6n - 1$. Substituting $s_n = an + b$ into the recurrence relation for $n \geq 2$ gives

$$an + b = [a(n-1) + b] + 6[a(n-2) + b] + 6n - 1.$$

Therefore $an + b = (7a + 6)n + (7b - 3a - 1)$, and hence $a = 7a + 6$ and $b = 7b - 13a - 1$. It follows that $-6 = 6a$ and $13a + 1 = 6b$, so that $a = -1$ and $b = -2$. An easy calculation shows that $an + b = -n - 2$ does satsify the recurrence relation.

(b)   By Exercise 30, we must find a solution of the corresponding homogeneous second-order linear difference equation $v_n = v_{n-1} + 6v_{n-2}$. Then a solution to the given inhomogeneous equation will have the form $s_n = u_n + v_n$, where $u_n = -n - 2$ from (a). Now, $u_0 = -2$ and $u_1 = -3$, and so $v_0 = s_0 - u_0 = -4$ and $v_1 = s_1 - u_1 = 13$. The auxiliary equation for $v - n$ is $x^2 = x + 6$, which has the solutions -2 and 3. Thus we look for a solution of the form $v_n = c_1(-2)^n + c_2(3^n)$. Using the initial conditions, we see that

$$c_1 + c_2 = v_0 = -4$$
$$-2c_1 + 3c_2 = v_1 = 13.$$

Hence $c_1 = -5$ and $c_2 = 1$. Thus the solution to the homogeneous equation is $v_n = -5(-2)^n + 3^n$, and so $s_n = -5(-2)^n + 3^n - n - 2$ is the solution to the inhomogeneous equation.

33.   The binary search algorithm proceeds as follows.

b	e	m	a_m	Is $a_m = 6$?
1	4	$\lfloor \frac{1+4}{2} \rfloor = 2$	4	no; $a_m < 6$
4	4	$\lfloor \frac{3+4}{2} \rfloor = 3$	6	yes

35.   Let A denote the list consisting of 45 and 57 and B denote the list consisting of 59 and 87. To create the merged list C, we compare 45 with 59. Because $45 < 59$, we choose 45 to begin C. Next we compare 57 with 59 and choose 57 to be the second item in C. Because all of A has been copied into C, we complete C by copying each of the items from B into C. Thus 2 comparisons are needed to merge A and B into C.

37.   Let A denote the list consisting of 1, 3, 5, and 7 and B denote the list consisting of 2, 4, 6, and 8. To create the merged list C, we compare 1 with 2. Because $1 < 2$, we choose 1 to begin C. Next we compare 3 with 2 and choose 2 to be the second item in C. To determine the third item in C, we compare 3 with 4 and choose 3 to be the third item in C. Continuing in this manner, we compare 5 with 4, 5 with 6, 7 with 6, and 7 with 8. In all, 7 comparisons are needed to merge A and B into C.

39.   Consider any two ordered lists such that the last element of the first list is less than the first element of the second list, for example, the ordered lists $1, 2, \ldots, m$ and $m + 1, m + 2, \ldots, m + n$. In such a case, we compare each item in the first list with only the first item in the second list. Thus m comparisons are needed to merge the two lists into one.

41. Substituting from the recurrence relation, we have

$$S = 1 + 2s_0 x + 2s_1 x^2 + 2s_2 x^3 + \cdots = 1 + 2xS,$$

so $(1 - 2x)S = 1$. This leads to

$$S = \frac{1}{1 - 2x} = 1 + 2x + 4x^2 + \cdots + 2^n x^n + \cdots.$$

43. Substituting from the recurrence relation, we have

$$S = 1 + x + (-2s_1 - s_0)x^2 + (-2s_2 - s_1)x^3 + (-2s_3 - s_2)x^4 + \cdots$$
$$= 1 + x - 2x(s_1 x + s_2 x^2 + \cdots) - x^2(s_0 + s_1 x + s_2 x^2 + \cdots)$$
$$= 1 + x - 2x(S - 1) - x^2 S$$

Thus $S(1 + 2x + x^2) = 1 + 3x$, and so $S = (1 + 3x)/(1 + 2x + x^2) = (1 + 3x)/(1 + x)^2$.

45. We have

$$S = \frac{5}{1 + 2x} = 5[1 - (-2x)]^{-1} = 5[1 + (-2x) + (-2x)^2 + (-2x)^3 + \cdots].$$

Examining the pattern, we find that the general coefficient is $s_n = 5(-2)^n$ for all $n \geq 0$.

47. We have

$$S = \frac{1}{1 - 2x} + \frac{2}{1 - 3x} = 1(1 + 2x + 4x^2 + \cdots) + 2(1 + 3x + 9x^2 + \cdots).$$

Examining the pattern, we find that $s_n = 2^n + 2(3^n)$ for all $n \geq 0$.

49. The value of a_r is the coefficient of the x^r term in

$$(1 + x + x^2 + x^3)(1 + x + x^2)(1 + x + x^2 + x^3 + x^4 + x^5).$$

51. Since each winner gets either 2, 3, or 4 prizes, the value of a_r is the coefficient of the x^r term in $(x^2 + x^3 + x^4)^6$.

53. The value of a_r is the coefficient of the x^r term in

$$(1 + x^5 + x^{10} + x^{15} + \cdots)(1 + x^{12} + x^{24} + \cdots)(1 + x^{25} + x^{50} + \cdots).$$

Here the power of x chosen from the first factor corresponds to the amount paid for 5-cent stamps, etc.

55. The value of a_r is the coefficient of the x^r term in

$$(1 + x + x^2 + x^3 + \cdots)(1 + x^5 + x^{10} + \cdots)(1 + x^{10} + x^{20} + x^{30} + \cdots).$$

Chapter 10

Combinatorial Circuits and Finite State Machines

10.1 LOGICAL GATES

1. The Boolean expression is $(x \wedge y) \vee x$.

3. The Boolean expression is $((x' \vee y) \wedge x)'$.

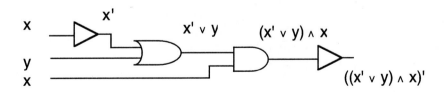

5. The Boolean expression is $(x'' \vee y') \wedge x'$.

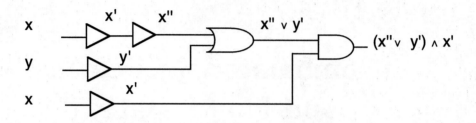

7. The Boolean expression is $(x' \wedge (y' \wedge x))'$.

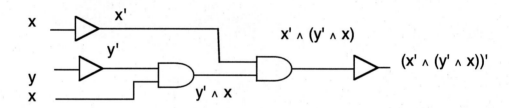

9. See the answer in the back of the text.

11. See the answer in the back of the text.

13. See the answer in the back of the text.

15. $(1 \vee 1) \wedge (1' \vee 0) = 1 \wedge (0 \vee 0) = 1 \wedge 0 = 0$

17. $(0 \wedge (1 \wedge 0))' = (0 \wedge 0)' = 0' = 1$

19.

x	y	x'	$x \wedge y$	$x' \wedge y$	$(x \wedge y) \vee (x' \wedge y)$
0	0	1	0	1	1
0	1	1	0	1	1
1	0	0	0	0	0
1	1	0	1	1	1

21.

x	y	z	x'	z'	$x \wedge z'$	$x' \vee y$	$(x \wedge z') \wedge (x' \vee y)$
0	0	0	1	1	0	1	0
0	0	1	1	0	0	1	0
0	1	0	1	1	0	1	0
0	1	1	1	0	0	1	0
1	0	0	0	1	1	0	0
1	0	1	0	0	0	0	0
1	1	0	0	1	1	1	1
1	1	1	0	0	0	1	0

23.

x	y	x'	$y \vee x'$	$x \wedge (y \vee x')$
0	0	1	1	0
0	1	1	1	0
1	0	0	0	0
1	1	0	1	1

25.

x	y	x'	y'	$x \wedge y$	$x' \wedge y'$	$(x' \wedge y')'$	$(x \wedge y) \vee (x' \wedge y')'$
0	0	1	1	0	1	0	0
0	1	1	0	0	0	1	1
1	0	0	1	0	0	1	1
1	1	0	0	1	0	1	1

27.

x	y	z	y'	z'	$x \vee y'$	$x \wedge z'$	$(x \vee y') \vee (x \wedge z')$
0	0	0	1	1	1	0	1
0	0	1	1	0	1	0	1
0	1	0	0	1	0	0	0
0	1	1	0	0	0	0	0
1	0	0	1	1	1	1	1
1	0	1	1	0	1	0	1
1	1	0	0	1	1	1	1
1	1	1	0	0	1	0	1

29.

x	y	$(x \vee y) \wedge x$	$x \vee (y \wedge x)$
0	0	0	0
0	1	0	0
1	0	1	1
1	1	1	1

equivalent

217

31.

x	y	$(x \wedge y)'$	$x' \vee y'$
0	0	1	1
0	1	1	1
1	0	1	1
1	1	0	0

equivalent

33.

x	y	$(x \wedge y) \vee (x \vee y)$	$x \vee y$
0	0	0	0
0	1	1	1
1	0	1	1
1	1	1	1

equivalent

35.

x	y	$x' \vee y'$	$((x \wedge y) \vee (x' \wedge y)) \vee (x' \wedge y')$
0	0	1	1
0	1	1	1
1	0	1	0
1	1	0	1

not equivalent

37.

x	y	$x \vee (x \wedge y)$
0	0	0
0	1	0
1	0	1
1	1	1

equivalent

39.

x	y	z	$((x \vee y) \wedge (x' \vee y)) \wedge (x \vee z)$	$(x \vee y) \wedge (x' \vee z)$
0	0	0	0	0
0	0	1	0	0
0	1	0	0	1
0	1	1	1	1
1	0	0	0	0
1	0	1	0	1
1	1	0	1	0
1	1	1	1	1

not equivalent

218

41.

x	y	z	$y' \wedge (y \vee z')$	$y' \wedge x'$
0	0	0	1	1
0	0	1	0	1
0	1	0	0	0
0	1	1	0	0
1	0	0	1	0
1	0	1	0	0
1	1	0	0	0
1	1	1	0	0

not equivalent

43. Let the Boolean variables w, d, and s be 1 when the window signal is heard, door is open, and safety switch is thrown, respectively. The alarm should sound when $w \vee (d \wedge s') = 1$. The circuit is shown below.

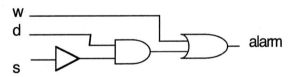

45. Let the Boolean variables be $x_1, x_2, \ldots, x_n$. Let X, Y, and Z be Boolean expressions in these variables. Then X has the same truth table as X. Also if X and Y have the same truth table, then so do Y and X. Finally, if X and Y have the same truth table and Y and Z have the same truth table, then X and Z have the same truth table.

47. If $x = 0$ the output of the AND-gate is 0 and so the output of the circuit is 1. If $x = 1$ the output of the AND-gate is the same as its lower input. But this input is the opposite of the output of the AND-gate. This is contradictory, and the output is undefined.

10.2   CREATING COMBINATORIAL CIRCUITS

1.

x	y	$x \wedge y$	$y \wedge x$
0	0	0	0
0	1	0	0
1	0	0	0
1	1	1	1

3.

x	$x \wedge x$
0	0
1	1

5.

x	y	x'	y'	$x \wedge y$	$(x \wedge y)'$	$x' \vee y'$
0	0	1	1	0	1	1
0	1	1	0	0	1	1
1	0	0	1	0	1	1
1	1	0	0	1	0	0

7.

x	x'	$x \wedge x'$
0	0	0
1	1	0

9.

$$
\begin{aligned}
(x \wedge y) \vee (x \wedge y') &= x \wedge (y \vee y') && \text{(by (c))} \\
&= x \wedge 1 && \text{(by (f))} \\
&= x && \text{(by (g))}
\end{aligned}
$$

11.

$$
\begin{aligned}
x \wedge (x' \vee y) &= (x \wedge x') \vee (x \wedge y) && \text{(by (c))} \\
&= 0 \vee (x \wedge y) && \text{(by (f))} \\
&= x \wedge y && \text{(by (g))}
\end{aligned}
$$

13.

$$
\begin{aligned}
(x' \vee y)' \vee (x \wedge y') &= (x'' \wedge y') \vee (x \wedge y') && \text{(by (j))} \\
&= (x \wedge y') \vee (x \wedge y') && \text{(by (i))} \\
&= x \wedge y' && \text{(by (e))}
\end{aligned}
$$

15.

$$
\begin{aligned}
(x \wedge y)' \vee z &= (x' \vee y') \vee z && \text{(by (j))} \\
&= x' \vee (y' \vee z) && \text{(by (b))}
\end{aligned}
$$

17.

$$
\begin{aligned}
(x \wedge y) \wedge ((x \wedge w) \vee (y \wedge z)) &= (((x \wedge y) \wedge (x \wedge w)) \vee ((x \wedge y) \wedge (y \wedge z)) && \text{(by (c))} \\
&= ((y \wedge x) \wedge (x \wedge w)) \vee ((x \wedge y) \wedge (y \wedge z)) && \text{(by (a))} \\
&= (y \wedge (x \wedge (x \wedge w))) \vee (x \wedge (y \wedge (y \wedge z))) && \text{(by (b) twice)} \\
&= (y \wedge ((x \wedge x) \wedge w)) \vee (x \wedge ((y \wedge y) \wedge z)) && \text{(by (b) twice)} \\
&= (y \wedge (x \wedge w)) \vee (x \wedge (y \wedge z)) && \text{(by (e) twice)} \\
&= ((y \wedge x) \wedge w) \vee ((x \wedge y) \wedge z) && \text{(by (b) twice)} \\
&= ((x \wedge y) \wedge w) \vee ((x \wedge y) \wedge z) && \text{(by (a))} \\
&= (x \wedge y) \wedge (w \vee z) && \text{(by (c))}
\end{aligned}
$$

220

19.

x	y	z	$x \wedge (y \vee z)$	$(x \wedge y) \vee z$
0	0	0	0	0
0	0	1	0	1*
0	1	0	0	0
0	1	1	0	1*
1	0	0	0	0
1	0	1	1	1
1	1	0	1	1
1	1	1	1	1

Output is different in starred rows.

21.

x	y	z	$(x \wedge y) \vee (x' \wedge z)$	$(x \vee x') \wedge (y \vee z)$
0	0	0	0	0
0	0	1	1	1
0	1	0	0	1*
0	1	1	1	1
1	0	0	0	0
1	0	1	0	1*
1	1	0	1	1
1	1	1	1	1

Output is different in starred rows.

23. $(x' \wedge y) \vee (x \wedge y')$

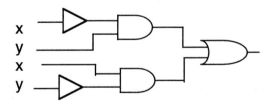

25. $(x' \wedge y \wedge z) \vee (x \wedge y' \wedge z) \vee (x \wedge y \wedge z')$

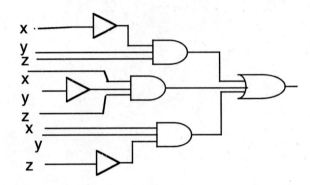

27. $(x' \wedge y \wedge z) \vee (x \wedge y' \wedge z) \vee (x \wedge y \wedge z)$

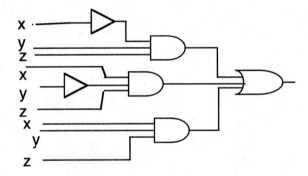

29. There are 9 gates shown. Each AND-gate with 3 inputs represents two original gates, so a total of $9 + 3 = 12$ are needed.

31. As in Exercise 29 a total of $6 + 3 = 9$ gates are needed.

33. As in Exercise 29 a total of $8 + 6 = 14$ gates are needed.

35.

x	y	$x \rightarrow y$	$\sim (x \rightarrow y)$
0	0	1	0
0	1	1	0
1	0	0	1
1	1	1	0

The corresponding minterm is $x \wedge y'$, and the circuit is as follows.

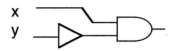

37. Translating the conditions into symbols gives $(a \wedge b \wedge c') \vee ((b \vee c) \wedge d') \vee (a \wedge d)$.

39. All rules hold. See Theorems 2.1 and 2.2.

41. We have $a \wedge (0 \vee a/) = a \wedge a' = 0$ by properties (v) and (vii). But also $a \wedge (0 \vee a') = (a \wedge 0) \vee (a \wedge a') = (a \wedge 0) \vee 0 = a \wedge 0$ by properties (iv), (vii), and (v). Likewise $1 = a \vee a' = a \vee (1 \wedge a') = (a \vee 1) \wedge ((a \vee a') = (a \vee 1) \wedge 1 = a \vee 1$ by properties (vii), (v), (iv), (vii), and (v).

43. We have $(a \vee a) \wedge (a \vee a') = (a \vee a) \wedge 1 = a \vee a$ by properties (vii) and (v). Also $(a \vee a) \wedge (a \vee a') = a \vee (a \wedge a') = a \vee 0 = a$ by properties (iv), (vii), and (v). Likewise $(a \wedge a) \vee (a \wedge a') = (a \wedge a) \vee 0 = a \wedge a$ by properties (vii) and (v). Also $(a \wedge a) \vee (a \wedge a') = a \wedge (a \vee a') = a \wedge 1 = a$ by properties (iv), (vii), and (v).

45. We have $(a \vee b) \vee (a' \wedge b') = (b \vee a) \vee (a' \wedge b') = b \vee (a \vee (a' \wedge b')) = b \vee ((a \vee a') \wedge (a \vee b')) = b \vee (1 \wedge (a \vee b')) = b \vee (a \vee b') = b \vee (b' \vee a) = (b \vee b') \vee a = 1 \vee a = 1$ by properties (ii), (iii), (iv), (vii), (v), (ii), (iii), and (vii), and Exercise 41. Likewise $(a \vee b) \wedge (a' \wedge b') = (a \vee b) \wedge (b' \wedge a') = ((a \vee b) \wedge b') \wedge a' = ((a \wedge b') \vee (b \wedge b')) \wedge a' = ((a \wedge b') \vee 0) \wedge a' = (a \wedge b') \wedge a' = (b' \wedge a) \wedge a' = b' \wedge (a \wedge a;) = b' \wedge 0 = 0$ by exactly the same sequence of reasons. Then $(a \vee b)' = a' \wedge b'$ by Exercise 40. A similar proof shows that $(a \wedge b)' = a' \vee b'$.

10.3   KARNAUGH MAPS

1. $(x' \wedge y') \vee (x' \wedge y) \vee (x \wedge y)$

3. $(x' \wedge y' \wedge z') \vee (x' \wedge y' \wedge z) \vee (x' \wedge y \wedge z') \vee (x' \wedge y \wedge z)$

5. $(w' \wedge x' \wedge y' \wedge z') \vee (w' \wedge x' \wedge y \wedge z') \vee (w' \wedge x \wedge y' \wedge z') \vee (w' \wedge x \wedge y \wedge z') \vee (w \wedge x \wedge y' \wedge z)$

7. $x \vee y'$

9. $y \vee x' \vee (y' \wedge z')$

11. $(x \wedge z) \vee (w' \wedge x) \vee (w \wedge x' \wedge y') \vee (w' \wedge y \wedge z')$

13. The Karnaugh map is shown below.

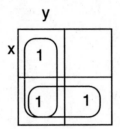

15. The Karnaugh map is shown below.

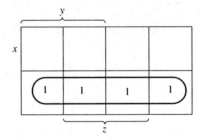

17. The Karnaugh map is shown below.

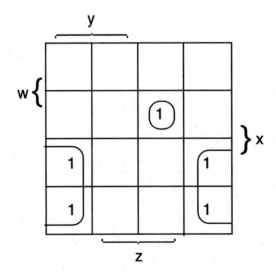

19. The Karnaugh map is shown below.

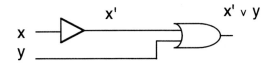

21. The Karnaugh map is shown below.

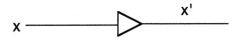

23. The Karnaugh map is shown below.

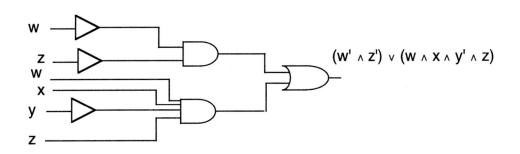

25. $(x' \wedge z) \vee (y' \wedge z)$

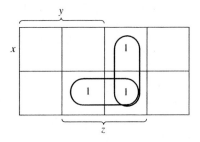

27. $(x' \wedge z) \vee (x \wedge y' \wedge z')$

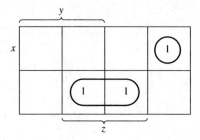

29. $x \vee (y' \wedge z')$

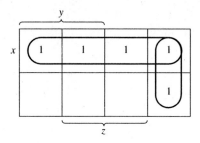

31. $(w \wedge x) \vee (y' \wedge z')$

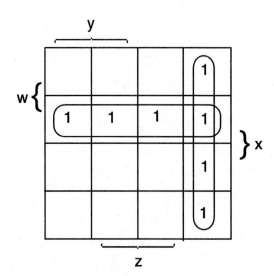

33. $(x' \land y' \land z) \lor (w \land x' \land y')$

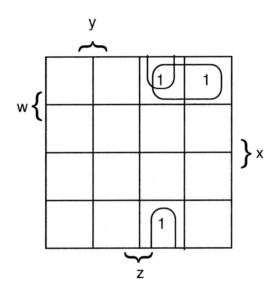

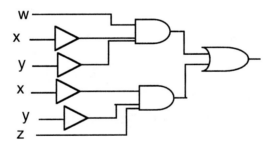

35. There are 4 two-cell horizontal groups in each of the 4 rows, and 4 two-cell vertical groups in each of the 4 columns, for 32 in all.

37. If $n = 1$ what we need to prove is just the definition of $x_1 \lor x_2 \lor \ldots \lor x_m \lor y_1 \lor \ldots y_n$. Suppose $(x_1 \lor \ldots \lor x_m) \lor (y_1 \lor \ldots \lor y_n) = x_1 \lor x_2 \lor \ldots \lor x_m \lor y_1 \lor \ldots y_n$. Then

$(x_1 \lor \ldots \lor x_m) \lor (y_1 \lor \ldots \lor y_n \lor y_{n+1})$

$$\begin{aligned}
&= (x_1 \lor \ldots \lor x_m) \lor ((y_1 \lor \ldots \lor y_n) \lor y_{n+1}) && \text{(definition)} \\
&= ((x_1 \lor \ldots \lor x_m) \lor (y_1 \lor \ldots \lor y_n)) \lor y_{n+1} && \text{(Theorem 10.1)} \\
&= (x_1 \lor \ldots \lor x_m \lor y_1 \lor \ldots \lor y_n) \lor y_{n+1} && \text{(induction hypothesis)} \\
&= x_1 \lor \ldots \lor x_m \lor y_1 \lor \ldots \lor y_n \lor y_{n+1} && \text{(definition)}.
\end{aligned}$$

Thus by the principle of mathematical induction the equation is true for all positive integers n.

227

39. If $n = 1$ the result is obvious. Suppose $(x_1 \vee x_2 \vee \ldots \vee x_n)' = x_1' \wedge x_2' \wedge \ldots \wedge x_n'$. Then

$$
\begin{aligned}
(x_1 \vee \ldots \vee x_{n+1})' &= ((x_1 \vee \ldots \vee x_n) \vee x_{n+1})' && \text{(definition)} \\
&= (x_1 \vee \ldots \vee x_n)' \wedge x_{n+1}' && \text{(Theorem 10.1)} \\
&= (x_1' \wedge \ldots \wedge x_n') \wedge x_{n+1}' && \text{(induction hypothesis)} \\
&= x_1' \wedge \ldots \wedge x_{n+1}' && \text{(definition)}.
\end{aligned}
$$

Thus by the principle of mathematical induction the equation is true for all positive integers n.

41. An expression counted by r_n has $n-1$ symbols $\vee$, and so $n-1$ expressions $A \vee B$. Each of these yields 4 expressions counted by r_{n+1} formed by substituting $(x_{n+1} \vee A) \vee B$, $(A \vee x_{n+1}) \vee B$, $A \vee (x_{n+1} \vee B)$, and $A \vee (B \vee x_{n+1})$ for a total of $4(n-1)r_n$ expressions. If C is counted by r_n, we also have $x_{n+1} \vee C$ and $C \vee x_{n+1}$ counted by r_{n+1}. Thus $r_{n+1} = 4(n-1)r_n + 2r_n = (4n-2)r_n$.

10.4 FINITE STATE MACHINES

1. See the answer in the back of the text.

3. See the answer in the back of the text.

5. See the answer in the back of the text.

7. See the answer in the back of the text.

9. See the answer in the back of the text.

11.

Input	start	1	0	1	1	0	0	1
State	x	x	y	x	x	y	z	y

13.

Input	start	y	x	x	x	y
State	B	C	C	C	C	A

15.

Input	start	x	y	z	x	y	z	x
State	A	B	A	B	A	A	B	A

Input string accepted.

17.

Input	start	x	y	x	x	y	y
State	B	C	A	B	C	A	C

Input string not accepted.

19. See the answer in the back of the text.

21. See the answer in the back of the text.

23. See the answer in the back of the text.

25. See the answer in the back of the text.

27.

Input	start	2	1	0	1	2	1	1
State	A	A	B	A	B	B	B	B
Output		y	w	y	w	w	x	x

29.

Input	start	3	2	2	1	1	3
State	A	C	A	C	C	C	C
Output		y	z	z	y	y	z

31. We start in state a, but move to the accepting state b and stay there whenever a 1 is input. Thus the machine accepts exactly those input strings containing a 1.

33. We start in state a, and move one state clockwise whenever a 1 is input. Since there are three states, we will be in the accepting state b after $1, 4, 7, \ldots$ symbols 1 are input. Thus a string with n symbols 1 is accepted exactly when $n \equiv 1 \pmod 3$.

35. See the answer in the back of the text.

37. See the answer in the back of the text.

39. Clearly the relation of equivalence is reflexive and symmetric. Suppose F is equivalent to G and G is equivalent to H. Then a string is accepted by F if and only if it is accepted by G, and this happens if and only if it is accepted by H. Thus F and H are equivalent.

SUPPLEMENTARY EXERCISES

1. The Boolean expression is $(x' \vee y) \wedge (z \wedge x)'$.

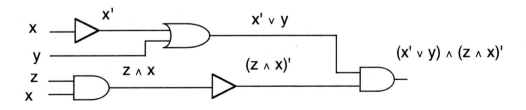

x	y	z	x'	$x' \vee y$	$z \wedge x$	$(z \wedge x)'$	$(x' \vee y) \wedge (z \wedge x)'$
0	0	0	1	1	0	1	1
0	0	1	1	1	0	1	1
0	1	0	1	1	0	1	1
0	1	1	1	1	0	1	1
1	0	0	0	0	0	1	0
1	0	1	0	0	1	0	0
1	1	0	0	1	0	1	1
1	1	1	0	1	1	0	0

3. (a) $x \wedge (y \wedge z')' = x \wedge (y' \vee z'') = x \wedge (y' \vee z) = (x \wedge y') \vee (x \wedge z)$ by parts (j), (i), and (c) of Theorem 9.1.

   (b) When $x = y = z = 1$, then $x \wedge (y' \wedge z)' = 1 \wedge (0 \vee 1)' = 1 \wedge 1' = 1 \wedge 0 = 0$, while $(x \wedge y) \vee (x \wedge z') = (1 \wedge 1) \vee (1 \wedge 0) = 1 \vee 0 = 1$. Thus the expressions are not equivalent.

5. See the answers in the back of the text.

7. See the answer in the back of the text for the Boolean expression and circuit. The circuit has 10 gates, including 4 AND-gates with $3 = 2 + 1$ inputs, and 1 OR-gate with $4 = 2 + 2$ inputs. Thus to make the circuit with gates having 1 or 2 inputs would require $10 + 4(1) + 1(2) = 16$ gates.

9. See the answers in the back of the text.

11. The Karnaugh map is shown below.

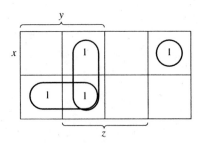

See the answer in the back of the text for the simplified Boolean expression and circuit. The new circuit has 7 gates, including 2 gates with $3 = 2 + 1$ inputs. Thus to make the circuit with gates having 1 or 2 inputs would require $7 + 2(1) = 9$ gates.

13. The Karnaugh map is shown below.

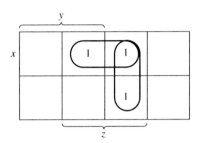

The simplified expression is $(x \wedge z) \vee (y' \wedge z)$. See the new circuit in the back of the text.

15.

Input	start	green	red	green	red	yellow
State	A	A	B	B	B	C

17.

Input	start	5	5	3	3	5	5	3
State	z	y	z	x	y	z	y	y

19.

Input	start	a	b	a	a	b	b	a
State	cold	cold	hot	hot	hot	cold	hot	hot
Output		-1	1	1	1	0	1	1

21.

Input	start	1	0	0	1	0	0	1
State	1	cider	x	x	1	x	x	1
Output			b	b	foam	b	a	foam	b

Appendix A

An Introduction to Logic and Proof

A.1 STATEMENTS AND CONNECTIVES

1. false statement (Florida is south of Georgia.)

3. true statement

5. not a statement (This sentence is neither true nor false.)

7. false statement (The simple statement "seven is an even integer" is false, and thus the entire conjunction is false.)

9. not a statement (This sentence is neither true nor false.)

11. true statement (The first simple statement is true.)

13. The negation is: $4 + 5 \neq 9$.

15. The negation is: California is the largest state in the United States.

17. In order to negate a sentence containing the word "all," change "all" to "some . . . not." Thus the negation is: Some birds cannot fly.

19. In order to negate a sentence containing the words "there is," change "there is" to "no . . . ." Thus the negation is: No man weighs 400 pounds.

21. The negation is: No students do not pass calculus. This negation would probably be most often written as: All students pass calculus.

23. The negation is: Not everyone enjoys cherry pie. (The negation could also be writtten as: Someone does not enjoy cherry pie.)

25. (a) The conjunction is: One is an even integer and nine is a positive integer. The statement is false, because at least one (in this case both) of the simple statements is false.

(b) The disjunction is: One is an even integer or nine is a positive integer. The statement is true because at least one (in this case the second) of the simple statements is true.

27. (a) The conjunction is: The Atlantic is an ocean and the Nile is a river. The conjunction is true.

(b) The disjunction is: The Atlantic is an ocean or the Nile is a river. The disjunction is true.

29. (a) The conjunction is: Birds have four legs and rabbits have wings. The conjunction is false, because at least one (in this case both) of the simple statements are false.

(b) The disjunction is: Birds have four legs or rabbits have wings. The disjunction is false, because both simple statements are false.

31. (a) The conjunction is: Flutes are wind instruments and timpani are string instruments. The conjunction is false, because at least one (in this case the second) of the simple statements is false.

(b) The disjunction is: Flutes are wind instruments or timpani are string instruments. The disjunction is true, because at least one (in this case the first) of the simple statements is true.

33. (a) The converse is: If I go to the movies, then this is Friday.

(b) The inverse is: If this isn't Friday, then I won't go to the movies.

(c) The contrapositive is: If I don't go to the movies, then this isn't Friday.

35. (a) The converse is: If Kennedy runs for president, then he won't run for the Senate.

(b) The inverse is: If Kennedy runs for the Senate, then he won't run for president.

(c) If Kennedy doesn't run for president, then he is running for the Senate.

A.2   LOGICAL EQUIVALENCE

1. The truth table is as follows.

p	q	$p \vee q$	$p \wedge q$	$\sim(p \wedge q)$	$(p \vee q) \wedge [\sim(p \wedge q)]$
T	T	T	T	F	F
T	F	T	F	T	T
F	T	T	F	T	T
F	F	F	F	T	F

3. The truth table is as follows.

p	q	$\sim p$	$p \vee q$	$\sim p \wedge q$	$(p \vee q) \rightarrow (\sim p \wedge q)$
T	T	F	T	F	F
T	F	F	T	F	F
F	T	T	T	T	T
F	F	T	F	F	T

233

5. The truth table is as follows.

p	q	r	$p \to q$	$p \vee r$	$(p \to q) \to (p \vee r)$
T	T	T	T	T	T
T	T	F	T	T	T
T	F	T	F	T	T
T	F	F	F	T	T
F	T	T	T	T	T
F	T	F	T	F	F
F	F	T	T	T	T
F	F	F	T	F	F

7. The truth table is as follows.

p	q	r	$\sim p$	$\sim q$	$\sim q \vee r$	$\sim p \wedge q$	$(\sim q \wedge r) \leftrightarrow (\sim p \vee q)$
T	T	T	F	F	F	T	F
T	T	F	F	F	F	T	F
T	F	T	F	T	T	F	F
T	F	F	F	T	F	F	T
F	T	T	T	F	F	T	F
F	T	F	T	F	F	T	F
F	F	T	T	T	T	T	T
F	F	F	T	T	F	T	F

9. The truth table is as follows.

p	q	r	$p \vee q$	$p \wedge r$	$(p \vee q) \wedge r$	$(p \wedge r) \vee q$	$[(p \vee q) \wedge r] \to [(p \wedge r) \vee q]$
T	T	T	T	T	T	T	T
T	T	F	T	F	F	T	T
T	F	T	T	T	T	T	T
T	F	F	T	F	F	F	T
F	T	T	T	F	T	T	T
F	T	F	T	F	F	T	T
F	F	T	F	F	F	F	T
F	F	F	F	F	F	F	T

11. The truth table for $\sim p \vee p$ is as follows.

p	$\sim p$	$\sim p \vee p$
T	F	T
F	T	T

Because each entry in the column corresponding to $\sim p \vee p$ is T, this statement is a tautology.

13. The truth table for $(\sim p \wedge q) \to \sim (q \to p)$ is as follows.

p	q	$\sim p \wedge q$	$q \to p$	$\sim (q \to p)$	$(\sim p \wedge q) \to \sim (q \to p)$
T	T	F	T	F	T
T	F	F	T	F	T
F	T	T	F	T	T
F	F	F	T	F	T

Because each entry in the column corresponding to $(\sim p \wedge q) \to \sim (q \to p)$ is T, this statement is a tautology.

15. The truth table for $\sim [((p \to q) \wedge (\sim q \vee r)] \wedge (\sim r \wedge p)]$ is as follows.

p	q	r	$p \to q$	$\sim q \vee r$	$\sim r \wedge p$	$\sim [((p \to q) \wedge (\sim q \vee r)] \wedge (\sim r \wedge p)]$
T	T	T	T	T	F	T
T	T	F	T	F	T	T
T	F	T	F	T	F	T
T	F	F	F	T	T	T
F	T	T	T	T	F	T
F	T	F	T	F	F	T
F	F	T	T	T	F	T
F	F	F	T	T	F	T

Because each entry in the column corresponding to $\sim [((p \to q) \wedge (\sim q \vee r)] \wedge (\sim r \wedge p)]$ is T, this statement is a tautology.

17. We proceed as in Example A.8 by making a truth table containing columns for each of the given statements. Such a table is shown below.

p	$\sim p$	$\sim (\sim p)$
T	F	T
F	T	F

Because the entries in the columns corresponding to p and $\sim (\sim p)$ are the same, these statements are logically equivalent.

19. A truth table containing columns for each of the given statements is shown below.

p	q	$p \to q$	$\sim (p \to q)$	$\sim q$	$p \vee q$	$\sim q \wedge (p \vee q)$
T	T	T	F	F	T	F
T	F	F	T	T	T	T
F	T	T	F	F	T	F
F	F	T	F	T	F	F

Because the entries in the columns corresponding to $\sim(p \to q)$ and $\sim q \wedge (p \vee q)$ are the same, these statements are logically equivalent.

21. Proceed as in Exercise 19 by making a truth table containing columns corresponding to the statements $p \to (q \to r)$ and $(p \wedge q) \to r$. If the 8 rows are arranged as in Exercise 15, then the entries in the columns corresponding to each of the given statements are T, F, T, T, T, T, T, T.

23. Proceed as in Exercise 19 by making a truth table containing columns corresponding to the statements $(p \vee q) \to r$ and $(p \to r) \wedge (q \to r)$. If the 8 rows are arranged as in Exercise 15, then the entries in the columns corresponding to each of the given statements are T, F, T, F, T, F, T, T.

25. **(a)** If the table containing the verification of statement (a) is arranged with 4 rows as in Exercise 1, then the entries in the column corresponding to each of the given statements are T, F, F, F. Thus the two statements are logically equivalent.
   (b) If the table containing the verification of statement (b) is arranged with 4 rows as in Exercise 1, then the entries in the column corresponding to each of the given statements are T, T, T, F. Thus the two statements are logically equivalent.

27. **(e)** If the table containing the verification of statement (e) is arranged with 8 rows as in Exercise 5, then the entries in the column corresponding to each of the given statements are T, T, T, T, T, F, F, F. Thus the two statements are logically equivalent.
   (f) If the table containing the verification of statement (f) is arranged with 8 rows as in Exercise 5, then the entries in the column corresponding to each of the given statements are T, T, T, F, F, F, F, F. Thus the two statements are logically equivalent.

29. **(i)** If the table containing the verification of statement (i) is arranged with 4 rows as in Exercise 1, then the entries in the column corresponding to each of the given statements are T, F, T, T. Thus the two statements are logically equivalent.

31. If the table containing the verification of modus ponens is arranged with 4 rows as in Exercise 1, then the entries in the column corresponding to the statement are T, T, T, T. Thus, the statement is a tautology.

33. **(a)** A truth table for "exclusive or" is shown below.

p	q	$p \veebar q$
T	T	F
T	F	T
F	T	T
F	F	F

(b)   A truth table for $\sim(p \leftrightarrow q)$ is shown below.

p	q	$p \leftrightarrow q$	$\sim(p \leftrightarrow q)$
T	T	T	F
T	F	F	T
F	T	F	T
F	F	T	F

Because the entries in the column corresponding to $\sim(p \leftrightarrow q)$ are the same as those in the column corresponding to the exclusive form of "or" in (a), $p \veebar q$ is logically equivalent to $\sim(p \leftrightarrow q)$.

A.3   METHODS OF PROOF

1. A truth table containing columns corresponding to the statements $\sim(p \wedge \sim q)$ and $p \rightarrow q$ is shown below.

p	q	$\sim q$	$p \wedge \sim q$	$\sim(p \wedge \sim q)$	$p \rightarrow q$
T	T	F	F	T	T
T	F	T	T	F	F
F	T	F	F	T	T
F	F	T	F	T	T

Since the corresponding entries in the last two columns are identical, the statements $\sim(p \wedge \sim q)$ and $p \rightarrow q$ are logically equivalent.

3. Let m be an integer. We will prove the contrapositive, which is the statement: If m is not odd, then m^2 is not odd. This is equivalent to the statement: If m is even, then m^2 is even. But this is the statement proved in Example A.9. Hence the original statement ("If m is an integer and m^2 is odd, then m is odd") is true as well.

5. Suppose that a divides b. Then $b = ka$ for some integer k. Hence for any positive integer c, $bc = (ka)c = k(ac)$. Thus ac divides bc.

7. Suppose that a divides b and b divides c. Then $b = ma$ for some integer m and $nb = c$ for some integer n. Hence $c = nb = n(ma) = (nm)a$. Thus a divides c because nm is an integer.

9. Suppose that p is prime, so that $p > 1$. Because q is a prime, the only divisors of q are 1 and q. Thus if p divides q, we must have $p = q$.

11. We prove the contrapositive of the given statement. Recall from A.1(g) that the negation of a conjunction is found by using DeMorgan's Laws. This gives the contrapositive statement as: If x is odd and y is odd, then xy is odd. Suppose that both x and y

237

are odd. Then there are integers m and n such that $s = 2m + 1$ and $y = 2n + 1$. It follows that

$$xy = (2m + 1)(2n + 1) = 4mn + 2m + 2n + 1 = 2(smn + m + n) + 1.$$

Hence xy is odd. Thus the contrapositive of the original statement is proved, and so the original statement is established. Thus if xy is even, then x is even or y is even.

13. The statement is false. For example, $3 + 5 = 8$. In fact, the sum of any two odd integers is always even.

15. The statement is false. For example, if $a = 3, b = 2$, and $c = 0$, then $ac = bc$, but $a \neq b$.

17. The statement "If 6 divides xy, then 6 divides x or 6 divides y" is false. For example, 6 divides $3 \cdot 2$, but 6 divides neither 3 nor 2.

19. Let a and b be odd. By Exercise 4, both a^2 and b^2 are odd. It follows that the sum of these two odd integers is even.

21. The statement is true. Suppose that n is odd. Then $n = 2k + 1$ for some integer k. Hence $n^2 - 1 = (2k + 1)^2 - 1 = 4k^2 + 4k = 4k(k + 1)$. But k and $k + 1$ are consecutive integers; so one of them must be even. Thus $k(k + 1)$ is even, and so $k(k + 1) = 2m$ for some integer m. Therefore $n^2 - 1 = 4k(k + 1) = 4(2m) = 8m$, and so $n^2 - 1$ is divisible by 8.

Conversely, suppose that 8 divides $n^2 - 1$. Then there exists an integer m such that $n^2 - 1 = 8m$. Thus $n^2 = 8m - 1$ is odd. By Example A.12, it follows that n is also odd.

23. The statement is false. Consider the case when $n = 41$:

$$(41)^2 + 41 + 41 = 41(41 + 1 + 1) = 41(43),$$

so that the number is composite when $n = 41$.

25. The statement is true. Using the division algorithm to divide n by 3, we may write n in the form $n = 3q + r$, where q and r are integers and $0 \leq r < 3$. Thus $r = 0, r = 1$, or $r = 2$.

Case 1: $r = 0$
In this case,
$$n^2 - 2 = (3q)^2 - 2 = 9q^2 - 2 = 3(3q^2) - 2,$$

which is not divisible by 3 because 3 does not divide -2.

Case 2: $r = 1$
In this case,

$$n^2 - 2 = (3q + 1)^2 - 2 = 9q^2 + 6q - 1 = 3(3q^2 + 2q) - 1,$$

which is not divisible by 3 because 3 does not divide -1.

Case 3: $r = 2$
In this case,

$$n^2 - 2 = (3q + 2)^2 - 2 = 9q^2 + 12q + 2 = 3(3q^2 + 4q) + 2,$$

which is not divisible by 3 because 3 does not divide 2.

Hence, for each positive integer n, $n^2 - 2$ is not divisible by 3.

27. The statement is true. Let p be a prime positive integer. In order to obtain a contradiction, we will assume that $\log_{10} p = \frac{m}{n}$, where m and n are positive integers. (Note that the integers m and n can be chosen to be positive because $\log_{10} p > 0$.) Then $p = 10^{m/n}$, and so $p^n = 10^m = 2^m \cdot 5^m$. Now, p must be even by Exercise 11, and thus $p = 2$. If $n \le m$, then $1 = 2^{m-n}5^m$, a contradiction. And if $n > m$, then $2^{n-m} = 5^m$, a contradiction since the left side is even and the right side is odd. Hence $\log_{10} p$ is not a rational number.

SUPPLEMENTARY EXERCISES

1. true statement

3. not a statement (There is no way to determine if this sentence is true nor false.)

5. false statement ($5^2 < 5!$.)

7. true statement

9. The negation is: There exists a square that is a triangle. (false)

11. The negation is: No scientist from the United States has received a Nobel prize. (false)

13. The negation is: $2 + 2 \le 4$ and 1 is not a root of $x^5 + 1 = 0$. (true)

15. The negation is: In circling the globe along a line of latitude, one must not cross the equator exactly twice or not cross the North Pole or not cross the South Pole. (true)

17. (a) Squares have four sides and triangles have three sides. (true)
(b) Squares have four sides or triangles have three sides. (true)

19. (a) If $3 > 2$, then $3 \times 0 > 2 \times 0$; and if $4 = 5$, then $5 = 9$. (false)
(b) If $3 > 2$, then $3 \times 0 > 2 \times 0$; or if $4 = 5$, then $5 = 9$. (true)

21. (a) If $3^2 = 6$, then $3 + 3 = 6$. (true)
(b) If $3 + 3 \ne 6$, then $3^2 \ne 6$. (true)

   (c) If $3^2 \neq 6$, then $3 + 3 \neq 6$. (false)

23. **(a)** If $3 \times 2 = 6$, then $3^2 = 6$. (false)
   (b) If $3^2 \neq 6$, then $3 \times 2 \neq 6$. (false)
   (c) If $3 \times 2 \neq 6$, then $3^2 \neq 6$. (true)

25. The truth table is as follows.

p	q	$(p \vee q) \wedge \sim p$	$\sim[(p \vee q) \wedge \sim p]$	$\sim[(p \vee q) \wedge \sim p] \wedge \sim p$
T	T	F	T	F
T	F	F	T	F
F	T	T	F	F
F	F	F	T	T

27. The truth table is as follows.

p	q	r	$[p \wedge (r \wedge (\sim p \vee q))] \rightarrow [(p \wedge r) \wedge (\sim p \vee q)]$
T	T	T	T
T	T	F	T
T	F	T	T
T	F	F	T
F	T	T	T
F	T	F	T
F	F	T	T
F	F	F	T

29. The statement is a tautology.

31. The statement is not a tautology—it is false whenever p is false and q is true.

33. The statements are logically equivalent.

35. The statements are logically equivalent.

37. The statement is false. For example, 6 cannot be written in this fashion.

39. Assume that x is a perfect square and also that x is even. Because x is a perfect square, $x = m^2$ for some positive integer m. Since $x = m^2$ is even, then the result established in Example A.12 shows that m is even. Hence $m = 2k$ for some positive integer k, and so $x = m^2 = (2k)^2 = 4k^2$. Since k^2 is a positive integer, x is divisible by 4.

41. We prove this statement by mathematical induction. The statement is true for $n = 1$ because $1^5 - 1 = 0$ is divisible by 10. Suppose that 10 divides $k^5 - k$ for some positive integer k. The number $(k+1)^5 - (k+1)$ expands to

$$(k+1)^5 - (k+1) = (k^5 + 5k^4 + 10k^3 + 10k^2 + 5k + 1) - (k+1)$$
$$= (k^5 - k) + (5k^4 + 10k^3 + 10k^2 + 5k)$$
$$= (k^5 - k) + 5k(k+1)(k^2 + k + 1).$$

By the induction hypothesis, 10 divides $k^5 - k$, and 10 divides $5k(k+1)(k^2 + k + 1)$ because either k or $k+1$ is even. Thus 10 divides $(k+1)^5 - (k+1)$, and so the given statement is true for all positive integers by the principle of mathematical induction.

43. Consider the consecutive integers $k+1$ and k. The difference of their cubes is

$$(k+1)^3 - k^3 = (k^3 + 3k^2 + 3k + 1) - k^3 = 3k^2 + 3k + 1 = 3k(k+1) + 1.$$

Because either k or $k+1$ is even, the expression $3k(k+1)$ is always even, and so the difference of these cubes is odd.

Appendix B

Matrices

1. Only matrices of the same size can be added. Because A is a 2×3 matrix and B is a 3×2 matrix, the sum $A + B$ is not defined.

3. Because C and A are both 2×3 matrices, the sum $C + A$ is defined. To compute $C + A$, we add corresponding entries of C and A. Thus

$$C + A = \begin{bmatrix} -1 & 0 & 2 \\ 3 & 1 & 4 \end{bmatrix} + \begin{bmatrix} 1 & 2 & 3 \\ 0 & -1 & 2 \end{bmatrix}$$

$$= \begin{bmatrix} -1+1 & 0+2 & 2+3 \\ 3+0 & 1+(-1) & 4+2 \end{bmatrix}$$

$$= \begin{bmatrix} 0 & 2 & 5 \\ 3 & 0 & 6 \end{bmatrix}.$$

5. Recall that the product AB is defined when the number of columns in A equals the number of rows in B. Because A is a 2×3 matrix and B is a 3×2 matrix, the product AB is defined. Moreover, the product of an $m \times n$ matrix and an $n \times p$ matrix is an $m \times p$ matrix; so AB is a 2×2 matrix. The row i, column j entry of AB is the sum of the products of corresponding entries from row i of A and column j of B. Hence

$$AB = \begin{bmatrix} 1 & 2 & 3 \\ 0 & -1 & 2 \end{bmatrix} \begin{bmatrix} -2 & 0 \\ 1 & 3 \\ 4 & 1 \end{bmatrix}$$

$$= \begin{bmatrix} 1(-2) + 2(1) + 3(4) & 1(0) + 2(3) + 3(1) \\ 0(-2) + (-1)(1) + 2(4) & 0(0) + (-1)(3) + 2(1) \end{bmatrix}$$

$$= \begin{bmatrix} 12 & 9 \\ 7 & -1 \end{bmatrix}.$$

7. Because the number of columns in A, which is 3, does not equal the number of rows in B, which is 2, the product AC is not defined.

9. Proceeding as in Exercise 5, we obtain

$$AB = \begin{bmatrix} 1 & 0 \\ -1 & 1 \end{bmatrix} \begin{bmatrix} 3 & 2 \\ 1 & 5 \end{bmatrix} = \begin{bmatrix} 1(3) + 0(1) & 1(2) + 0(5) \\ (-1)(3) + 1(1) & (-1)(2) + 1(5) \end{bmatrix} = \begin{bmatrix} 3 & 2 \\ -2 & 3 \end{bmatrix}.$$

11. Proceeding as in Exercise 5, we obtain

$$A^2 = \begin{bmatrix} 1 & 0 \\ -1 & 1 \end{bmatrix} \begin{bmatrix} 1 & 0 \\ -1 & 1 \end{bmatrix} = \begin{bmatrix} 1(1) + 0(-1) & 1(0) + 0(1) \\ -1(1) + 1(-1) & (-1)(0) + 1(1) \end{bmatrix} = \begin{bmatrix} 1 & 0 \\ -2 & 1 \end{bmatrix}.$$

13. Using the results of Exercises 11 and 12, we have

$$A^2B^2 = \begin{bmatrix} 1 & 0 \\ -1 & 1 \end{bmatrix} \begin{bmatrix} 11 & 16 \\ 8 & 27 \end{bmatrix} = \begin{bmatrix} 1(11) + 0(8) & 1(16)(0(27) \\ -2(11) + 1(8) & (-2)(16) + 1(27) \end{bmatrix} = \begin{bmatrix} 11 & 16 \\ -14 & -5 \end{bmatrix}.$$

15. Using the result of Exercise 11, we have

$$A^3 = AA^2 = \begin{bmatrix} 1 & 0 \\ -1 & 1 \end{bmatrix} \begin{bmatrix} 1 & 0 \\ -2 & 1 \end{bmatrix} = \begin{bmatrix} 1(1) + 0(-2) & 1(0) + 0(1) \\ -1(1) + 1(-2) & (-1)(0) + 1(1) \end{bmatrix} = \begin{bmatrix} 1 & 0 \\ -3 & 1 \end{bmatrix}.$$

17. Because A and B are both 3×3 matrices, the product AB is defined and is also a 3×3 matrix. Proceeding as in Exercise 5, we have

$$AB = \begin{bmatrix} 1 & 1 & 1 \\ 1 & 1 & 1 \\ 1 & 1 & 1 \end{bmatrix} \begin{bmatrix} 1 & 0 & 0 \\ 0 & -1 & 0 \\ 0 & 0 & 1 \end{bmatrix}$$

$$= \begin{bmatrix} 1(1) + 1(0) + 1(0) & 1(0) + 1(-1) + 1(0) & 1(0) + 1(0) + 1(1) \\ 1(1) + 1(0) + 1(0) & 1(0) + 1(-1) + 1(0) & 1(0) + 1(0) + 1(1) \\ 1(1) + 1(0) + 1(0) & 1(0) + 1(-1) + 1(0) & 1(0) + 1(0) + 1(1) \end{bmatrix}$$

$$= \begin{bmatrix} 1 & -1 & 1 \\ 1 & -1 & 1 \\ 1 & -1 & 1 \end{bmatrix}.$$

19. As in Exercise 17, we have

$$A^2 = \begin{bmatrix} 1 & 1 & 1 \\ 1 & 1 & 1 \\ 1 & 1 & 1 \end{bmatrix} \begin{bmatrix} 1 & 1 & 1 \\ 1 & 1 & 1 \\ 1 & 1 & 1 \end{bmatrix}$$

$$= \begin{bmatrix} 1(1) + 1(1) + 1(1) & 1(1) + 1(1) + 1(1) & 1(1) + 1(1) + 1(1) \\ 1(1) + 1(1) + 1(1) & 1(1) + 1(1) + 1(1) & 1(1) + 1(1) + 1(1) \\ 1(1) + 1(1) + 1(1) & 1(1) + 1(1) + 1(1) & 1(1) + 1(1) + 1(1) \end{bmatrix}$$

$$= \begin{bmatrix} 3 & 3 & 3 \\ 3 & 3 & 3 \\ 3 & 3 & 3 \end{bmatrix}.$$

21. Using the results of Exercises 19 and 20, we have

$$A^2B^2 = \begin{bmatrix} 3 & 3 & 3 \\ 3 & 3 & 3 \\ 3 & 3 & 3 \end{bmatrix} \begin{bmatrix} 1 & 0 & 0 \\ 0 & 1 & 0 \\ 0 & 0 & 1 \end{bmatrix}$$

$$= \begin{bmatrix} 3(1)+3(0)+3(0) & 3(0)+3(1)+3(0) & 3(0)+3(0)+3(1) \\ 3(1)+3(0)+3(0) & 3(0)+3(1)+3(0) & 3(0)+3(0)+3(1) \\ 3(1)+3(0)+3(0) & 3(0)+3(1)+3(0) & 3(0)+3(0)+3(1) \end{bmatrix}$$

$$= \begin{bmatrix} 3 & 3 & 3 \\ 3 & 3 & 3 \\ 3 & 3 & 3 \end{bmatrix}.$$

23. Using the result of Exercise 19, we have

$$A^3 = AA^2 = \begin{bmatrix} 1 & 1 & 1 \\ 1 & 1 & 1 \\ 1 & 1 & 1 \end{bmatrix} \begin{bmatrix} 3 & 3 & 3 \\ 3 & 3 & 3 \\ 3 & 3 & 3 \end{bmatrix}$$

$$= \begin{bmatrix} 1(3)+1(3)+1(3) & 1(3)+1(3)+1(3) & 1(3)+1(3)+1(3) \\ 1(3)+1(3)+1(3) & 1(3)+1(3)+1(3) & 1(3)+1(3)+1(3) \\ 1(3)+1(3)+1(3) & 1(3)+1(3)+1(3) & 1(3)+1(3)+1(3) \end{bmatrix}$$

$$= \begin{bmatrix} 9 & 9 & 9 \\ 9 & 9 & 9 \\ 9 & 9 & 9 \end{bmatrix}.$$

25. (a) We have

$$AB = \begin{bmatrix} 1 & -1 \\ -1 & 1 \end{bmatrix} \begin{bmatrix} 2 & 1 \\ 2 & 1 \end{bmatrix} = \begin{bmatrix} 1(2)+(-1)(2) & 1(1)+(-1)(1) \\ -1(2)+1(2) & -1(1)+1(1) \end{bmatrix} = \begin{bmatrix} 0 & 0 \\ 0 & 0 \end{bmatrix}$$

and

$$BA = \begin{bmatrix} 2 & 1 \\ 2 & 1 \end{bmatrix} \begin{bmatrix} 1 & -1 \\ -1 & 1 \end{bmatrix} = \begin{bmatrix} 2(1)+1(-1) & 2(-1)+1(1) \\ 2(1)+1(-1) & 2(-1)+1(1) \end{bmatrix} = \begin{bmatrix} 1 & -1 \\ 1 & -1 \end{bmatrix}.$$

(b) The calculation in (a) shows that matrix multiplication is not commutative, that is, AB need not equal AB. Also, the calculation of AB shows that the product of two matrices that are not zero matrices may equal a zero matrix.

27. Both $A+B$ and $B+C$ are defined and are $m \times n$ matrices. Thus both $(A+B)+C$ and $A+(B+C)$ are defined and are $m \times n$ matrices. Hence, to prove that $(A+B)+C = A+(B+C)$, we need only show that the corresponding entries of $(A+B)+C$ and $A+(B+C)$ are equal. Consider the row i, column j entries, where $1 \le i \le m$ and $1 \le j \le n$. If the row i, column j entries of A, B, and C are a, b, and c, respectively,